Augsburg College
George Sverdrup Library
Minneapolis, MN 55454

WITHDRAWN

D1030709

ARTIFICIAL INTELLIGENCE FOR SOCIETY

ARTIFICIAL INTELLIGENCE FOR SOCIETY

Edited by

KARAMJIT S. Gill

SEAKE Centre
Department of Computing
and Cybernetics
Brighton Polytechnic

JOHN WILEY & SONS

Chichester . New York . Brisbane . Toronto . Singapore

Copyright © 1986 by John Wiley & Sons Ltd

All rights reserved.

No part of this book may be reproduced by any means, or transmitted, or translated into a machine language without the written permission of the publisher.

Library of Congress Cataloging in Publication Data:

Main entry under title:

Artificial intelligence for society.

  Includes index.
  1. Artificial intelligence—Addresses, essays,
lectures. 2. Artificial intelligence—Social aspects—
Addresses, essays, lectures. I. Gill, Karamjit S.
Q335.5.A785 1986   006.3   85–29596

ISBN 0 471 90930 0

British Library Cataloguing in Publication Data:

Artificial intelligence for society.
  1. Artificial intelligence
  I. Gill, Karamjit S.
  006.3   Q335

ISBN 0 471 90930 0

Printed and bound in Great Britain

φ
335.5
A785
1986

Contents

168005

List of Contributors

Z. M. ALBES
Principal Consultant Psychologist
and
Founder of Downslands College
10 Beverley Court
South Cliff
Eastbourne
East Sussex
UK

DANILO BERTASIO
University of Genoa
Italy

P. M. DAVIES
City of Birmingham Polytechnic
Birmingham
UK

ANGUS DOULTON
National Interactive Video Centre
27 Marylebone Road
London, W1N 2BA
UK

MICHAEL DUMMETT
New College
Oxford
UK

RICHARD ENNALS
Department of Computing
Imperial College of Science and
Technology
180 Queen's Gate
London, SW7 2BZ
UK

KARAMJIT S. GILL
SEAKE Centre
Department of Computing and
Cybernetics
Brighton Polytechnic
Moulsecoomb
Brighton, BN2 4GJ
UK

R. T. GRIFFIN
City of Birmingham Polytechnic
Birmingham
UK

STEPHEN GWYN JONES
SEAKE Centre
Department of Computing and
Cybernetics
Brighton Polytechnic
Moulsecoomb
Brighton, BN2 4GJ
UK

GRAHAM J. HOWARD
Department of Art
Faculty of Art and Design
Coventry Lanchester Polytechnic
Gosford Street
Coventry, CV1 5RZ
UK

URSULA HUWS
Freelance Writer, Journalist and
Researcher
20 Cannonbury Square
London, N12 2AL
UK

FARHAD JAHEDI
SEAKE Centre
Department of Computing and
Cybernetics
Brighton Polytechnic
Moulsecoomb
Brighton, BN2 4GJ
UK

PETER LARGE
Technology Editor, *The Guardian*
119 Farrington Road
London, EC1R 3ER
UK

CHRIS MELLISH
Cognitive Studies Programme
School of Social Sciences
University of Sussex
Falmer
Brighton, BN1 9QN
UK

PARTHA MITTER
School of African and Asian
Studies
University of Sussex
Falmer
Brighton, BN1 9QN
UK

AJIT NARAYANAN
Department of Computer Science
University of Exeter
Prince of Wales Road
Exeter, EX4 4PT
UK

MASSIMO NEGROTTI
Chair of Sociology of Knowledge
University of Genoa
Italy

DEREK PARTRIDGE
Computing Research Laboratory
New Mexico State University
Las Cruces
New Mexico 88003
USA

JOHN PICKERING
Department of Psychology
University of Warwick
Coventry, CV4 7AL
UK

JULIE C. RUTKOWSKA
Cognitive Studies Programme
School of Social Sciences
University of Sussex
Falmer
Brighton, BN1 9QN
UK

RAVI SINGH
Black Rod Interactive Services
3 Ways House
40–44 Oldstone Street
London, W1P 7EA
UK

DAVID SMITH
National Foundation for
Educational Research
The Mere
Upton Park
Slough, SL1 2DQ
UK

BRYAN SPIELMAN
SINTA Software
3 Barnsfield Gardens
Brighton, BN2 2HQ
UK

GEOFF STEVENS
School of Industrial and Business
Studies
University of Warwick
Coventry CV4 7AL
UK

BRIAN V. STREET
School of Social Sciences
University of Sussex
Falmer
Brighton, BN1 9QN
UK

S. K. TOOLE
City of Birmingham Polytechnic
Birmingham
UK

STEVE TORRANCE
Cognitive Studies Programme
School of Social Sciences
University of Sussex
Falmer
Brighton, BN1 9QN
UK

JANET VAUX
Editor, *Machine Intelligence News*
16 Dulwich Road
London
UK

GAJENDRA VERMA
Department of Education
University of Manchester
Manchester
UK

BLAY WHITBY
SEAKE Centre
Department of Computing and
Cybernetics
Brighton Polytechnic
Moulsecoomb
Brighton, BN2 4GJ
UK

M. J. WINFIELD
Department of Computing
City of Birmingham Polytechnic
Birmingham
UK

SHARON WOOD
Cognitive Studies Programme
School of Social Sciences
University of Sussex
Falmer
Brighton, BN1 9QN
UK

MASOUD YAZDANI
Department of Computer Science
University of Exeter
Prince of Wales Road
Exeter, EX4 4PT
UK

Foreword

Technology progressively transforms our lives; but so adaptable are we that we scarcely notice the process. We have come, for example, to take it for granted that we can transfer ourselves in an hour or two from Athens to London or to Luton, or, in a slightly longer but still essentially negligible time, from London to Katmandu; this is no source of wonder to us, but merely part of the unregarded scaffolding of our lives, as anaesthetics have long been and transplants are rapidly becoming. In the same way, our mode of existence is so dependent on the telephone that we cannot imagine conducting our affairs without it. This adaptability, a salient characteristic of our species, responsible for our success perhaps more than our intelligence, applies as much to prevalent attitudes as to material capabilities; it is the reason why a historical perspective is acquired only with such difficulty. Up to only a quarter of a century ago, for instance, it was an all but universal assumption that procured abortion was disgraceful, indeed wicked; yet, in a televised discussion between young people about Northern Ireland, one of them gave the illegality of abortion in the Republic as his objection to living there: 'I would not want to live anywhere where abortion was not allowed', he said, with the air of one mentioning something which, with life, liberty and the pursuit of happiness, was one of the acknowledged rights of man. But, though our attitudes change, leaving scarcely a trace behind, and though technology so rapidly and radically alters the circumstances of our lives, we ourselves change hardly at all, in our needs, our pains and pleasures, our reactions to one another: that is what makes much of the fiction of the past so readily intelligible to us, although the characters live their lives in a framework utterly different from our own.

As the contributors to this volume, and the participants in the conference which gave rise to it, are all vividly aware, most technological changes are prompted by motives unconnected with concern for their effect on people's lives: the fascination of scientists and technicians with intellectual problems, and their pride in solving them; the desire for commercial gain; the military need for more effective means of killing people, or for the means of killing them in greater numbers. By no conceivable standards, however, could the human race as a whole be judged to be doing well. In Africa, vast numbers have been facing starvation, which did not come upon them unexpectedly, but had been long predicted. Something was belatedly done to save some from death, though there is little concern about those who are again threatened; but few of the world's rulers display any urgent determination so to order things that a like disaster should not occur again, or to enquire into its underlying causes in the world economic system with a view to removing them. Millions live in destitution, or are subject to killings, torture and oppression; mankind, as a body, lacks the will to end these horrors.

Amazing technological advances thus take place against a background of widespread and unnecessary misery from which there appears little hope of relief; and those advances themselves perturb the conditions that bring about this misery. Almost any increase in human capabilities is a potential means for improving the human condition generally; but almost any perturbation in the existing order is likely, on balance, to worsen that condition, unless its introduction is accompanied by adjustments designed by those who have thought through its economic and social effects and have the overt intention to render the change beneficial to the human community as a whole. This is because the malign effects follow automatically, while the good ones must be planned; without such planning, only the wealthy and the powerful benefit. The full impact of information technology and artificial intelligence has not yet been felt; arguably, if we contrive to continue to avoid the further use of nuclear weapons, their effects will be far more far-reaching than any other technological innovation of the past three centuries. It is with the exploitation of the new technology to benefit society at large that the contributors are preoccupied. With the ability of artificial intelligence to provide useful models of, or illuminating insights into, human mental operations they are little concerned, and, to the extent that they consider it, they tend to scepticism; but the attention of most is directed, rather, to the applications of AI.

They are aware, too, that the mere sense of responsibility to society is not enough to forestall disaster. In the 1940s, the architects and town planners were thoroughly imbued with a feeling of social responsibility. They allowed their enthusiasm for the modern style and the modern techniques that made it possible to persuade them, and to persuade others, that they had the secret of constructing the ideal urban environment for twentieth-century people. We know now that they were creating, instead, little hells for the unfortunates who were relegated to them. Pride in one's own achievements may blind one to their real social effects, and will surely do so if one lacks the understanding of other people's needs that only close contact can give. If we are to avoid the disasters that revolutionary technological change can bring in its train, much deep thought will be needed by those not blinded by their own technical prowess. This book seeks to point the direction such thought must take. This it succeeds in doing, as will be obvious to anyone who registers how different is its tone from that of almost all else that is being written on this subject; and for this we must thank its authors and its editor, who was also the organizer of the conference that gave it birth, Dr Karamjit Gill.

Michael Dummett

Acknowledgements

The inspiration for organizing the AI For Society Conference and bringing together various contributions in this volume is a result of discussions with a large number of colleagues and friends. I am specially indebted to Michael Dummett for chairing the conference, to Surojit Sen, Farhad Jahedi and Stephen Jones for their active participation in organizational work on the conference, and to the participants who contributed to the conference and to this volume. Bill Tymms organized an exhibition on visual literacy, Blay Whitby and the CAAAT Project team provided general organizational and administrative support for the conference.

The volume could not have been completed without the enthusiastic support of three people—Surojit Sen and Satinder Gill who made helpful suggestions and constructive comments on numerous occasions, and Farhad Jahedi who helped in proof reading and in completing most of the administrative work.

Much needed support and encouragement for the debate on wider social applications and implications of new technology has come from various people especially Swasti Mitter, John Bishop, Philipa Kennedy, Karl Chidsey, David Smith, John Young, Martin Kender, Robert Upward, Richard Byrne, Terry Furber, Jonathan Seagrave, Ajit Naraynan, Masoud Yazdani, Derek Partridge, Sylvia Weir, Ranan Banerji, Richard Ennals, Margaret Boden, Mike Cooley, Robin Murray, John Burrow, Ray Miller and Jack Sutherland.

If this work makes any contribution, some credit should go to my wife Ajit Gill who has provided the moral and emotional support during and after the conference.

The SEAKE Centre is very grateful to Geoffrey R Hall, Director Brighton Polytechnic for his continuous support for the Centre's work.

On behalf of the SEAKE Centre, I gratefully acknowledge the sponsorship of the following organizations for the AI For Society Conference:
—Economic and Science Research Council
—South East Arts Council
—American Express
—Brighton Polytechnic

Karamjit S Gill

Introduction: What is Artificial Intelligence?

This volume is a selection from the Proceedings of the Third Annual Conference on *AI for Society* organized by the SEAKE Centre at Brighton Polytechnic. The SEAKE (Social and Educational Applications of Knowledge Engineering) Centre has been established in the Department of Computing and Cybernetics at the Polytechnic to research into the issues of knowledge transfer and interactive knowledge-based systems for education and training with particular emphasis on the needs of the disadvantaged.

Its two previous conferences brought together various groups concerned with the aim of introducing AI methodologies, their implications and applications. From our own experience of organizing these conferences and from subsequent discussions with interested groups and individuals, it became clear that there was a growing body of people actively engaged with the larger implications of AI developments and with the contributions these advances can make to meet the needs of society. It was in order to provide a forum for articulating these concerns and with the hope that the subsequent discussions would help to create an active community of commitment that it was proposed to hold a two-day conference to consider AI and IT developments within a framework of social, economic and cultural needs and responsibilities.

AIMS

Current AI and IT developments reflect an increasing preoccupation with task-specific technical skill acquisition and comparative neglect of the application of technology to enhance general human competence to handle the problems of a technological society. The Conference aimed to provide for alternative approaches to these developments. It brought together researchers, practitioners, decision makers and those involved in education, literacy, employment, training, health and welfare in the hope of providing a framework for discussion and interchange of ideas on these issues.

The Conference set itself four broad aims:

1. Critical examination of current or potential research and developments in the above areas
2. Introduction of AI methodologies and software tools for IKBS applications to educationalists, social workers and others who may be interested in applying them in their own fields
3. Discussion of the potential of AI research and consequent social, economic and political responsibilities
4. AI and IT in a cultural perspective: issues of cultural perception and understanding; multimedia interactive, learning environments for human cognitive development

The Conference, which was chaired by Professor Michael Dummett of the University of Oxford, brought together a distinguished body of practitioners and researchers. Some idea of the diversity of topics and approaches may be gained from the collection of papers in this volume. The volume has been planned to reflect the main areas of concern explored at the Conference.

The opening part, AI—Problems and Perspectives, attempts to draw a preliminary contour map of a complex and often confusing territory, where warnings and prophetic denunciations are readily audible but signposts to direct the unwary explorer are few and far between.

In the paper that inaugurated the Conference, Peter Large of *The Guardian* raises some issues of concern about the wider use of the so-called 'thinking' machine which is itself dependent upon two uncertainties: our knowledge of ourselves and our knowledge of our own creation. While acknowledging the contribution of AI techniques of expert systems in areas of comparatively 'firm' knowledge and experience such as those of medical diagnosis and training, and mineral exploration, he warns us of the dangers of reinforcing human biases and solidifying wide social and economic assumptions rooted in the past.

This opening address is followed by a paper by the Editor which raises issues concerning automation and knowledge transfer, which reflects the work being carried out at the SEAKE Centre. It argues that the automatic intelligent machine is primarily designed to meet the requirements of the machine-centred approaches to production and economic wealth creation. This limits its applications to those domains in which only knowledge of experts needs to be acquired and processed by the machine. These approaches mean that the user is merely a recipient rather than a participant in the acquisition and interpretation of knowledge, which is the central feature of human-centred approaches. If the knowledge-based technologies are to benefit all groups of society, their central concern must be with human development and social wealth creation. Consequently their design would necessitate taking into account not only knowledge of human needs but also of the relevant social and cultural contexts.

Richard Ennals points out the current emphasis of artificial intelligence research towards military applications and proposes an outline of a 'strategic health initiative' (SHI) as an alternative strategy for long-term research. He argues that such a strategy offers a way forward to exploring non-military applications of advanced technology for social needs and the civilian market, especially within the European context. He is optimistic about the commitment of some AI researchers for socially beneficial applications of AI. He warns us, however, that if funds are not committed to the civilian field in the United Kingdom, there will undoubtedly be economic pressure on these AI researchers to participate in military projects such as the 'Star Wars' project.

Derek Partridge's paper poses some fundamental challenges to current AI research practices. He considers issues of designing AI systems and points out the dangers of large-scale unreliable and incomprehensible computer programs which are increasingly being used to control a large part of our lives. He suggests that the complex and ill-structured nature of AI problem domains necessitates an incremental, evolutionary and exploratory process for designing AI systems. He proposes that AI programs should be validated on the basis of formal software engineering methods of specification and design. He also points out that AI problem domains are highly context sensitive, and hence the design of AI systems must deal with incomplete knowledge. This then requires incremental updating of knowledge as well. In other words, AI software should be capable of self-adaptation and self-modification which in turn requires the design of machine-learning programmes to deal with these issues. However, the current state of the art of machine learning does not give us the confidence to deal with the above issues of adaptability. He goes on to warn that the imperfect machine-learning methodologies could give rise to societal disruption.

The part on AI—Philosophical Issues contains papers by Ajit Narayanan, Steve Torrance, Janet Vaux and Julie Rutkowska. They succeed in uniting a wide-ranging enquiry into the philosophical foundations of AI with a rigorous and closely argued examination of the issues involved.

Ajit Narayanan provides an incisive analysis within which AI as a discipline may be judged as a science or a technology. He challenges the claim that AI is a science or a study of mental faculties through the use of computer models. He also considers the claims of many AI researchers that AI is making progress and is producing 'surprising' results. He points out that the criteria used for measuring this progress are dependent upon the progress of technologies such as computer architecture, not AI itself.

He proposes that if the progress of AI is dependent upon technology, then the best safeguard for AI's future lies in considering it as a technology rather than as a science. This view of AI would enable the public to focus on the 'accountability' and constraints of its applications in the real world.

Steve Torrance comments on the nature of intelligence in the context of present AI developments; he accepts the plausibility of the computer simulating human cognitive activities such as problem solving and task performing. He is, however, concerned about the further AI claims that any mental state, in principle, is computer simulable and therefore computationally explicable. He then goes into some fundamental philosophical and controversial issues concerning the nature of consciousness, ethics and moral responsibilities.

Janet Vaux comments on the failure of communication and a significant disparity between AI theorists and philosophers and wonders whether AI theorists are really interested in issues such as the mind–body problem and the 'other minds' problem in the same way as the philosophers are. She points out that there may be potential for the use of formal reasoning for automation and mechanical control, but AI as a model of the human being can run up against difficulties.

Julie Rutkowska's paper contains a productively sceptical scrutiny of the linkages between AI and philosophies of mind that are assumed, as a matter of course, by certain proponents of AI. She points out that human intelligence cannot be understood solely in terms of internal structures and processes. She argues that fundamental to human development are behavioural processes and physical and social environments—factors which AI has failed to take account of. She suggests that instead of emphasis on 'task-oriented learning', AI should emphasize 'knowledge-oriented learning'.

The part on AI—culture and the Arts may come as something of a surprise to readers of AI literature but it is, we should point out, an essential part of the shift in emphasis that we are adumbrating. Human needs and human learning necessarily take place in social–cultural configurations and as the whole history of recent educational reforming initiatives makes painfully clear, not to be sensitive to these contexts of learning and living is to put at risk the success of the enterprise itself.

Partha Mitter considers some of the underlying assumptions behind the optimism that new technology would benefit all societies. He points out that new technology is embedded in Western culture and hence AI shares a notion of cultural homogeneity common to much scientific thinking. He challenges the notion of homogeneity and its assumptions of universality of human behaviour and of needs of all societies. He emphasizes the need for taking into account both the variety and complexity of human cultural experience and the cultural dimension in designing appropriate technologies. He points out the consequent limitations and dangers of 'robotic thinking processes' and argues that the cultural factor cannot be ignored in the domain that actually deals with the acquisition of knowledge, since human values and judgements are embedded in the culture itself.

Massimo Negrotti and Danila Bertasio draw a distinction between the information processing capabilities of the 'thinking machine' and knowledge processing capabilities of humans. They point out that whilst the machine can theoretically achieve a very high level of formal reasoning, it is unable to reproduce knowledge. They consider the cultural roots of AI and emphasize that we should consider computer technology as culturally conditioned and constituted by the thought and cultural premises of the society it comes from.

Blay Whitby considers the cultural role of new technology and argues that it is predominantly based on military culture and is designed to fulfil the needs of that culture. He points out that the computer culture has had its greatest successes in areas where clear goals and the means of achieving them have been defined. In the same way AI culture has, so far, had its successes in areas such as game playing and theorem proving, which is consistent with the criteria of military culture. The paper shows a relationship between the terminology of AI and the language of dominant military culture.

Graham Howard considers issues of image representation and interpretation and outlines some of the dangers of automating these processes. He provides an insight into the nature of images and the way they have been used by the dominant discourse to propagate knowledge and belief structures. Since, in his view, the dominant discourse of society is naturally reproduced in the dominant technologies of that society, he is concerned that automation of image interpretation could lead to censorship and image cleansing which has enormous social and political implications. However, he sees a great potential in new technology for increasing our knowledge of the world and of ourselves through a reassessment of how images may be understood and used. He warns that such potential can only be realized if we recognize the power of images and question the role of the dominant discourse and dominant technology.

The next part on Social Issues—Realities and Aspirations lies at the very heart of concerns and approaches which a conference like ours must explore. Too often ignored or marginalized, these issues, together with the cultural concerns of the preceding part, nevertheless represent the points of development necessary to the long-term health and relevance of our own discipline. A whole cluster of related topics is at issue here, and it would be idle to pretend that anything approaching an exhaustive coverage of the related topics has been achieved. Nevertheless, we do think the topics selected for treatment have a central relevance to our own concerns and, in our opinion, deserve the attention of the AI community.

John Pickering and Geoff Stevens consider the potential and limitations of new technology for enhancing life chances of the handicapped and their control of their own environments. They point out some of the complexities involved in the use and application of new technology for meeting individual and group needs of the handicapped. As to the creation of new employment opportunities for the disabled to compete on an equal footing, they caution us against the danger of equipping the handicapped with skills which commit them to the least attractive and least rewarding jobs.

Gajendra Verma considers issues concerning the educational needs and occupational aspirations of ethnic minority pupils within the context of a multiracial, multicultural and multilingual society. He emphasizes that many of the disadvantages experienced by these pupils are a result of the nature of the educational system which reflects values and ideologies of the dominant culture. He suggests that since cultural pluralism is becoming a permanent feature of British society, there may be a role for new technology for enhancing the awareness and understanding of teachers, careers advisors and educationists about the nature of the social and cultural contexts within which the needs and aspirations of ethnic minority pupils could be meaningfully interpreted and catered for.

Ursula Huws considers the impact of technology on various aspects of women's lives, placing her analysis in the context of the pressures and distortions generated by structurally embedded factors such as prejudice and skill segregation both in the educational system and wider society.

Brian Street considers the analogy between conventional literacy and computer literacy. He points out that the concept of conventional literacy practices such as reading and writing are already imbedded in ideology and cannot be isolated or treated as 'neutral' or merely 'technical'. He challenges the notion that technology is 'neutral' and can be detached from the specific social and cultural contexts. He is hence opposed to the consideration of 'computer literacy' purely in terms of technological transfer and padagogic techniques without taking into account the ideological and cultural assumptions on which it is based.

Education has, of course, long been an area where AI has made substantial contributions and the papers in the part on AI, IT and Education examine in some depth various facets of the interface between AI and education.

David Smith's paper comments on the current debate on curriculum development and considers its implications for the future needs of society. He points out the educational poverty of present IT-based curriculum reform with its underlying economic utilitarianism. He questions the emphasis placed on mechanistic instruction and 'technological literacy'. He warns of the obsolescence built into the 'electronic sabre tooth' curriculum and points out that any curriculum which does not address issues of the needs and rights of citizens will fail and the price of failure could be social catastrophe. He puts forward a challenge to scientists and teachers to take up their responsibility of ensuring that their research does not occur in social and moral vacuum.

Masoud Yazdani's paper gives an overview of AI techniques and AI environments which, in his view, are beginning to influence the educational applications of computers. As an AI practitioner, he provides an optimistic view of the role of AI techniques such as machine learning in designing practical educational aids.

Chris Mellish's paper gives an introduction to logic programming and discusses some of the benefits of using logic programming languages such as Prolog for designing intelligent knowledge-based systems. The paper is relevant in this context because of the considerable interest in logic programming and logic programming languages among AI researchers and practitioners involved in expert systems work.

Brian Spielman gives a professional view of current practices used in the development of educational software and provides a practical guide to those interested in the design and use of computer software for education.

Angus Doulton points out the need for investigations into human learning processes and issues of interactivity, if we are to exploit the full potential of interactive video technology for education. He emphasizes the importance of the proper use of technology and of people for social applications and warns of the danger of being overinfluenced by the immense storage capabilities of the machine.

Ravi Singh gives a practitioner's viewpoint of the role of interactive video for designing learning systems for education and training. He considers issues of interactivity for courseware design and provides an overview of the differences between videotape and videodisc systems.

These investigations should be placed in the broader contexts of social and cultural imperatives sketched in the preceding two parts. The ever-accelerating pace of technological developments and the related growth of a large reservoir of deskilled youth and adults make it urgently necessary that we devise educational and training programmes relevant to a situation of rapidly changing needs and skills. At the same time we need to bear in mind what is, in fact, a constant theme running throughout these proceedings: knowledge transfer or the imparting of technological expertise cannot be conceived of as occurring in a vacuum. The relevant social and cultural coordinates must never be lost sight of if the training and educational initiatives are to bear fruit. The range of interrelated issues that such an approach brings into focus will form the basis of our next conference.

This volume concludes with a final part on a selected range of applications of new technology in specific areas.

Sharon Wood comments on the difficulty of formalizing knowledge of social situations. She emphasizes the need for designing expert systems which assist and guide the user rather than manufacturing solutions of problems and which do not reduce the autonomy of the individuals by excluding them from decision-making processes.

Stephen Jones emphasizes the need for taking into account the individual's life experiences and skills for designing expert systems for education and training of disadvantaged adults.

Michael Winfield et al. discuss the design of an expert system for

Euresis for use by the social workers and raise some of the design issues based on their experience of their pilot work.

Farhad Jahedi considers the design of an interactive learning aid for children with learning difficulties and underlines the utility of such aids as an integral part of an overall teaching programme.

Z. M. Albes provides a glimpse of the complexity of the nature of the speech handicap and gives a practical example of how speech technology can be used for enhancing the quality of life of the handicapped. She, however, emphasizes the need for designing those technologies which can be adapted to the needs of the human rather than the human having to adapt to the limitations of the machine.

The various groups whose concerns are addressed in this part are too often relegated to the periphery of mainstream AI research and developments. They are central to our own ongoing research initiatives and it is fitting that the volume should conclude with a part devoted to their needs.

Karamjit S. Gill
October 1985

PART 1

AI—Problems and Perspectives

Artificial Intelligence for Society
Edited by K. S. Gill
© 1986 John Wiley & Sons Ltd

1.   IS AI A NOTIFIABLE DISEASE?

PETER LARGE Technology Editor, *The Guardian*

The Australian comedian Bill Kerr had an act around the halls which began with the words: 'I don't want to worry you but. . . .' What followed was a deadpan recital of the perils the audience faced, from a collapse of the balcony to an inferno in the pit. This paper may sound a bit like that. As an amateur observer of the scene I want to raise some questions about the wider implications of the artificial intelligence technique of *expert systems*.

For the benefit of the uninitiated, this is the method whereby a computer can provide professional advice through gathering, codifying, then applying the accumulated knowledge and experience of the human specialist in whatever field, from orthopaedics to gardening. The computer, under the guidance of someone given the reassuring title of a knowledge engineer, builds the expert system by establishing perhaps hundreds of rules (if this, then that) in cross-questioning sessions with the human specialist.

In the mid-1970s, a group of British computer scientists, led by Professor Donald Michie at Edinburgh University, coined a better phrase for the process. They called it *knowledge refining*. Sadly, the American jargon of expert system has won the day.

Professor Michie claimed in 1978 that 'a reliability and competence of codification can be produced which far surpasses the highest level that the unaided human expert has ever, perhaps even could ever, attain'. So far that boast seems far from proven, but I am not going to present you with a Luddite bawl. There are already excellent examples of the value of expert systems in areas of comparatively firm human knowledge and experience—in medical diagnosis and training, in mineral exploration and appropriately in helping electronic engineers to solve the puzzle of a computer breakdown.

So-called expert systems are even available today for cheap home computers, offering help with tax returns. The danger with these simple systems is that they inevitably lack the depth whereby, if the computer puts an apparently bizarre question, you can ask the system to explain its reasons.

Much more exciting is the work now underway in mating expert systems with the interactive videodisc. Thereby, we will be able to cross-question a real person on the screen for help in, for example, untangling the arcane ways of government bureaucracy. Some of this work will be presented in later papers.

They may be all right but what is worrying is that—often through a moulding of expert systems with computer modelling—governments and business are already beginning to use so-called thinking machines which rely on two uncertain sciences: our knowledge of ourselves and our knowledge of our own creation, the computer.

By the 1990s, computers may well be widely employed in framing judgemental decisions, long before we have even sorted out (religion and philosophy apart) how our own minds work. Professor Joseph Weizenbaum, of the Massachusetts Institute of Technology, put the danger neatly in his book *Computer Power and Human Reason* (Weizenbaum, 1976). He called it the imperialism of instrumental reasoning and he warned against the danger of assuming that we could ever talk of such human concepts as risk, courage, trust or endurance in terms of a machine. When Weizenbaum wrote that in 1977 the argument was not much more than an academic nicety. Today it is real.

Again, it must be emphasized that this is not a Luddite yell to stop research and development in more certain areas, but there is a fear that the artificial intelligence people, who are now coming into their own after languishing in the background since the 1960s, may not be giving enough attention to the deeper implications of their work and to explaining themselves to the public.

There could be great temptations to leap before we can walk—particularly as competition becomes more intense between the United States, Japan and Europe to produce the fifth generation of much more powerful computers for the 1990s. We might repeat—with subtler and deeper dangers this time—the blunders of the 1960s when, as one pundit has put it, people were using the wrong computers, in the wrong ways, for the wrong reasons and for the wrong tasks.

A central problem already with us is this: who is the expert whose codified expertise we are relying on in an expert system?

In the case of the electronics engineer using an expert system to help in solving a computer breakdown, the problem is minimal: if it works, it works. In the case of medical diagnosis and training, things are more complicated, but still not too much of a problem. The student will have heard of specialist X and will know that X will have tested the system himself, having produced it in partnership with a knowledge engineer. However, it is a different matter when we come to systems supposed to help us in making business or negotiating decisions—systems that more truly try to adapt the mathematical logic of the computer to areas of subjective choice and human judgement, sometimes using controversial psychological theories to do so. How do we then trace and judge the assumptions behind those systems?

In 1983, Sir John Mason, head of the Meteorological Office, surprisingly felt it necessary in his presidential address to the British Association to make the obvious point that computer models used in economic forecasting tend to reflect their creators' personal and political judgements. He said society already depended on giant computer models whose accuracy and usefulness could not be assessed. Even in weather forecasting, where models are based on fundamental laws of physics, we do not yet know enough to be sure of them.

Those sorts of problems are also appearing at the individual level. The excellent work of Dr Karamjit Gill and his team at Brighton has demonstrated that. To quote briefly from their handbook:

> Current artificial intelligence and information technology developments reflect an increasing preoccupation with task-specific technical skill acquisition and comparative neglect of the applications of technology to enhance general human competence to handle the problems of a technological society.

Without going into the political implications of how those problems might be solved, one can see dangers that go one stage deeper in their individual impact. Some trade union research has shown that the use of computer-aided design can go beyond dehumanizing the designer's work environment: it can also restrict the options for creativity, because what is incarcerated in the officially approved software defines the borders of permitted thinking. Someone with a bright but maverick idea could not face an added bias against change—the bias engendered by the huge investment in that computer software itself as well as in the assumptions behind it.

Dr Mike Cooley, the former engineering union president who is a research fellow of the Open University, has drawn on his own experience in the aerospace industry to emphasise the need for a different pattern of development that would safeguard the intuitive leap to new solutions.

We could conclude that the wrong sort of use of expert systems might reinforce those dangers at the detailed level of work as well as helping to solidify wider social and economic assumptions rooted in the past. The use of artificial intelligence, as of information technology itself, is about power. Of course, that has always been the case. The value or otherwise of any technology depends on how we use it. It is another aspect of that hoary but sound adage: don't blame the computer, blame the people!

What is new about some merging uses of artificial intelligence is that we may be extending our still inadequate understanding of the impact of computers on society into the arena of our inadequate understanding of our own thought processes. How does one engineer thought? We do not have to fly into esoteric and pointless arguments about 'Can a machine grow consciousness?' before that question arises.

In case I have sounded too much like Bill Kerr, I will end with two sanguine observations. Two centuries on—and after much pain—we have finally got the first Industrial Revolution roughly right, at least in bringing vastly wider opportunities and wealth to the hundreds of millions lucky enough to live in the rich half of the world. Nearly 40 years on, we have even got routine uses of the computer roughly right—though we still do not fully understand how we design the things in the first place.

It will therefore probably work out roughly right in the end—so far as our human inadequacies and man's inhumanity to man allow. It would be nice if, this time around, we could get it right a bit sooner, so that our children reap the benefits, instead of our children and our grandchildren reaping the pain and our great grandchildren getting the benefits. The papers in this volume make some notable attempts to do that.

REFERENCES

Weizenbaum, J. (1976). *Computer Power and Human Reason: From Judgement to Calculation*. W.H. Freeman, San Francisco
SEAKE Centre. (1984). *Introduction*, SEAKE Centre, Brighton.

Artificial Intelligence for Society
Edited by K. S. Gill
© 1986 John Wiley & Sons Ltd

2. THE KNOWLEDGE-BASED MACHINE: ISSUES OF KNOWLEDGE TRANSFER

Karamjit S. Gill SEAKE Centre, Brighton Polytechnic

ABSTRACT

Recent developments in new technology, especially knowledge-based intelligent machines with capabilities of natural language interaction, oral communication and visual and automatic creation and use of knowledge, present a challenge to the human self as the sole creator of knowledge, on the one hand, and provide a challenge to its use for the benefit of human development, on the other. Machine intelligence technologies in the form of expert systems and intelligent robots are being used as a powerful focus for machine-centred approaches to economic wealth creation. This approach is now being challenged by a group of workers in the field who seek to emphasize human-centred approaches to social wealth creation. While this diversity of approaches is to be welcomed, there seems to be an overemphasis on the machine-centred notion of progress giving rise to the concept of the 'creative computer' as the indispensible centre of this progress. However, if society is to benefit from new technologies, then this pluralism must be based on a common concern for human develop-

7

ment. There is thus a consequent need for devising technologies which put human needs and skills at their very centre.

FOCUS OF CURRENT TECHNOLOGICAL DEVELOPMENTS

Recent developments in new technology, especially under the Fifth Generation Computer Systems[1] programmes such as the Alvey (UK) and the ESPRIT (EEC) provide us with a well-guided emphasis on the machine-centred approaches and their applications. This emphasis seems primarily based on the notion of economic wealth creation and the mechanistic criteria of efficiency, speed and certainty for production purposes.

Current work in areas such as robotics, CAD, speech recognition, vision, expert systems and front-end systems to a large extent contributes to and confirms the notion of wealth creation centred on the machine. Although people involved in these developments may, and some do, argue that ultimately humanity will benefit from these advances, there does not seem to be any serious indication of this concern being put into practice so far as human development and human welfare are concerned. On the other hand, there is increasing unemployment, deskilling, centralized control, rise in technical elitism and loss of individual liberties. It is this state of affairs, primarily engendered during the period of these technological advances, which prevents the general public, particularly the disadvantaged, from having faith in new technologies. However, this machine-centred approach is being increasingly challenged by researchers and as Cooley,[2] Rosenbrock,[3] Noble,[4] Galjaard[5] and Gill[6]. They emphasize human-centred approaches which consider human development, human creativity and human knowledge as the central basis for technological developments. Wealth creation in this context is considered to be dependent upon the advancement of human cognitive competences and human skills to cope with the technological society.

Michie and Johnston's recent book, *The Creative Computer*,[7] should provide a powerful instigation for modifying the machine-centred approaches so as to meet some of the major needs of society such as the eradication of poverty, hunger, disease and unemployment. If any of the orthodox AI researchers needed a lead given by an established AI authority, then Michie and Johnston have provided it when they say, 'The world is faced with a host of problems, great and small; among the great ones are overpopulation, poverty, disease, pollution, shortage of energy, international conflict and economic stagnation.'

Although the machine-centred approach to knowledge origination, generation and refinement stressed by Michie and Johnston is not in consonance with human-centred approaches, it is still a positive and encouraging contribution if it is going to lead to creating an environment in which human welfare is not considered subservient to material production. It is hoped that their emphasis on global human issues will at least encourage some AI researchers to participate in the development of human-centred technologies. In the human-centred approaches, computational and storage capabilities of the computer are seen to complement the knowledge and skills of the human. Cooley identifies the essence of this approach when he says:

> The computer and the human mind have different but complementary abilities. The computer excels in analysis and numerical computation, and human excels in pattern recognition, the assessment of complicated situations and intuitive leap to new situations. If these different abilities can be combined they will amount to something more powerful and more effective than anything we have seen before.[2]

Whatever approach we may elect to use, the central focus of socially useful technologies must be human welfare and social wealth. For these technologies to meet these concerns, their development must put human needs and skills at their very centre. The aim should be to design those technologies which enable the use of skills, ingenuity and the creativity of humans for enhancing their chances and opportunities in life. Such designs would necessarily require not only knowledge of human needs and expertise but also knowledge of the relevant social and cultural contexts. If we view the role of the computer in terms of the broader social and cultural contexts in which it is embedded, then a new order of priorities and consequently a new range of challenges begin to emerge. Their central concern should be with the application of new technology to meet human needs in areas such as education, employment, health, literacy, welfare and training. The SEAKE Centre document[8] summarizes some of the challenges and responsibilities offered by new technologies as follows:

1. The need to investigate the impact of new technology in the provision of access to education and training opportunities and the resultant unequal distribution of life chances
2. The provision of resources for socially useful research
3. Design and development of appropriate knowledge engineering tools to enhance the functional competence and enrich the quality of life of disadvantaged people

To meet the above challenges and responsibilities, there is an urgent need to shift resources from production-centred developments to those which make use of emerging knowledge-based technologies to meet wider human needs. In the design of such technologies, benefit and accessibility should be seen as inextricably intertwined. However, the development of such socially beneficial systems will depend upon our understanding of the nature of the new technology, the nature of human–machine interaction and the issues of knowledge transfer involved in this interaction.

NATURE OF TECHNOLOGY

To discuss the nature of new technology, we need briefly to consider the relationship between knowledge and technology and how this relationship has influenced the design of new knowledge-based technologies. The dialogue between Bruner and Miller[9] gives an insight into these issues in the light of an historical perspective. According to Bruner, Ernst Mach provided an analysis of the relationship between subjective and objective knowledge: 'All data, he argued, came out of experience, direct experience of the senses. The origins of knowledge, then, were subjective. From this subjective base we construct science.' According to Mach, one way of interpreting subjective experience is to project it outside; the other way is to project experience inside. In our present context, it is this notion of interpretation as a two-way process which is significant for understanding issues of human–machine interaction and knowledge transfer.

To understand AI's current emphasis on the automation of human knowledge processing, it is worth considering how shifts in technology have influenced the nature of the human–machine relationship over the last two centuries. The predominance of the nineteenth century's concern with energy gave rise to the consideration of the machine as essentially an extension of the arms and hands. As Bruner comments: 'It was the age of mechanics and energy transfer. Not surprisingly, the machine models of mechanics, the guiding metaphor of technology, was transferred to the human domain.' As technology shifted and the era of cybernetics dawned, information replaced energy as the central concern of the emerging technology. Interest in the notion of control and transformation of information provided the impetus for investigations into issues such as the human ability to know, limitations of human information processing, how we deal with uncertainty and how we use the feedback of our own actions to regulate these actions.

Primarily the focus of these investigations was still on the limitations of human control over the machine. However, the shift of technology from energy transfer to information transfer resulted in a new human–machine relationship. The machine was no longer an additional component of labour under direct human physical control; it was an independent tool under human mental control. Although the process of the separation of the machine from the human had started, the machine was still a model of human labour. Knowledge was embedded in humans and it was their experience which was central to the control and operation of increasingly complex and sophisticated machines. It was the collaboration of the machine and the knowledge and experience of the the artisan, crafts person, engineers and scientists which provided the basis for industrial expansion and economic prosperity.

The advent of the computer completed the process of separation of the machine from the human. The computer could code information provided to it, store it, manipulate it and decode the stored information.

The computer memory allowed for the separation of knowledge from the human and stored it in the form of software as part of the machine itself. The computer was now a complete functional machine with its own memory, information processing and control capabilities. Previous collaboration between the human and the machine was now transferred to collaboration between the machine and knowledge extracted from the human. Although the human was still responsible for software development, the separation process, however, opened the way for the design of the autonomous machine.

The computer not only produced a massive shift in technology but also a massive shift from human-centred production and control to machine-centred production and control. No longer was human labour, experience and expertise needed to drive the machine; it was intelligent software which was needed to provide these functions. The significance of the computer is much more than its functional capabilities; it lies in the development of its memory which enables storage, representation, organization, manipulation and translation of stored knowledge. It is these capabilities of the human domain which endow the computer with a sense of power and aliveness and consequently accounts for the awe in which it held by people. As Bruner indicates, it was this human aspect of the computer which led to the post-war investigations into such mentalistic matters as the selective representations of the world in memory, strategies for coordinating information, i.e. 'the cognitive revolution'. Simon's computer program. 'General Problem Solver',[10] embodied an effort to simulate human thinking and gave rise to the birth of artificial intelligence. This was the beginning of the present phase of new technology, which started the process of bringing together the physical and mental processes in the form an autonomous intelligent machine. Current developments in knowledge-based technologies such as expert systems and intelligent robots and their sponsorship under fifth generation initiatives, such as the Alvey and the ESPRIT programmes in Europe, show vividly society's investment in and commitment to the process of automation.

This emphasis on automation raises issues as to how the design of the intelligent machine would ensure dynamic human–machine interaction in the sense that:

1. There is a two-way transfer of knowledge between the human and the machine.
2. The interpretation of this knowledge takes place within given social and cultural contexts.

INTELLIGENT MACHINE: KNOWLEDGE TRANSFER

The design of an intelligent machine, however, requires the development of formal techniques of simulating human knowledge. In other words, the techniques should be so designed as to acquire knowledge from humans, represent it in the computer, manipulate it, modify and interpret it. For this to happen, what was needed was a 'model of knowledge' which the machine could simulate. However the only 'model of knowledge' which is available is itself human knowledge, which is not at all well understood. Human knowledge is subjective, imprecise, uncertain and fuzzy. To make matters worse for the believer of the autonomous machine, it embodies human characteristics such as individuality, creativity, distinctiveness, and it is embedded in the human's social and cultural forms. The view of human knowledge being sketched here is akin to that advanced by Polanyi in *Personal Knowledge* and other works. As a recent expositor of his thought has put it: 'All knowledge is personal, because only persons can know. The idea of an impersonal and therefore specially reliable knowledge, such as science is popularly supposed to offer, is absurd'.[11] In addition, the machine could only handle precise, certain, deterministic and rule-based objective knowledge.

The way this intractable knowledge transformational issue has so far been resolved, to a limited extent, is to strip human knowledge of all its human characteristics so as to objectify it in a form suitable for processing by the machine. Michie and Johnston's 'creative computer' perhaps represents the best of the intelligent machines centred on this notion of objectivity. This notion thus provides a basis for mechanizing and ultimately automating human knowledge. Assuming that Michie's notion of 'knowledge refining' enables AI researchers to turn human knowledge into a 'more precise, reliable and comprehensive' form, there still remains the problem of interpreting it both by the human and the machine cooperatively in varying social and cultural contexts and forms. The problem arises from the fact that knowledge stored in the machine has already been stripped of all human characteristics and contexts within which humans acquire and interpret knowledge meaningfully. On the other hand, humans still embody those intuitive and irrational traits which machines just cannot handle. In the context of our discussion on the nature of knowledge, the intelligent machine can only acquire, modify and reinterpret those forms of knowledge which are given to it. Thus issues of knowledge transfer and dynamic human–machine interaction remain outside the domain of the 'creative computer'. Michie and Johnston's use of 'creative' in this context must therefore be judged as somewhat Aesopian, for how anything can be creative when everything that we normally understand by this term is stripped from it is a perplexing question. It is thus clear that if we are to design intelligent knowledge-based environments for human development and socially useful purposes, then their design would need to take the following considerations into account:

1. The nature of knowledge acquisition. It is important that we conceive of knowledge acquisition in dynamic rather than static terms. In other words, the knowledge to be transmitted needs to be structured and sequenced in a manner that pays due heed to the generative principle

at the heart of all knowledge acquisition. The user needs not an inert prepackaged assortment of data but data so organized as to enable him to apply it in new contexts and thereby generate new knowledge.

2. The second consideration follows inevitably from the first. If knowledge acquisition is to have this inbuilt dynamic creativity—there must be a strong resonance between the mode of knowledge representation and the user's own social and cultural experiences. Knowledge acquisition does not take place in a vacuum, nor does it usually take place in those carefully disinfected laboratory environments so beloved by experimental scientists. It is rooted in the shifting, often confused, matrices of real life and is essentially interactive. Instead of deploring this and trying to devise ways around it, we should see in this a major opportunity and actively use the learner's context, his/her personal, cultural and social idiom.

3. The last consideration is linked to the argument outlined above, that knowledge acquisition is not only a dynamic process but an interactive one, utilizing the learner's own life experience and lifestyle. If we are to do this we find ourselves involved at once in issues of interpretation and translation. Michael Bakhtin, the Russian critic and thinker, argued that all communication, even in the same language, involves questions of translation. However that may be, it is certainly the case that knowledge transfer between different cultural groups must involve us in the joint activities of interpretation and translation. Cultural hermeneutics are part and parcel of our enterprise and we should not be overtimid in embracing it.

Hence issues of knowledge transfer may only be meaningfully investigated if social and cultural contexts are taken into account when designing knowledge-based technologies.

TECHNOLOGY AND HUMAN DEVELOPMENT

The autonomous intelligent machine is then appropriate for those domains which lend themselves to the formulation of precise, consistent and error-free knowledge and rule-based decision making in which only knowledge of experts needs to be acquired and processed by the machine and the user is only the recipient of knowledge, and which do not require dynamic human–machine interaction—in other words, domains in which there is only translation (one-way transfer) but no dynamic transfer of knowledge. From this standpoint, then, current design criteria of the intelligent machine leaves most of the human and social development domains outside its sphere of application. The AI researcher thus faces two alternatives: first to redesign the intelligent machine and focus on human-centred approaches of development and second to continue on the path of automation of human knowledge leading to the creation of autonomous machines. It seems that the established AI community is predominantly committed to the second alternative, which is perhaps not very surprising as this fits very well with the currently predominant view of production through automation.

To appreciate this focus, one needs only to look at AI projects and developments which are taking place at present in areas such as office automation, intelligent robot, speech recognition, computer vision, machine translation, machine learning and automated factory. When considering the significance of the Alvey programme for AI developments in the United Kingdom, it is worth noting that large-scale demonstrator projects include areas of product automation, planning and administration, office automation and machine–machine communication. All these developments are primarily concerned with: (a) industrial, commercial and military applications which lend themselves to mechanistic information processing, rule-based decision making and control and (b) the design of intelligent machines capable of automated knowledge processing, decision making and control.

These applications are not only consistent with the predominant notions of social progress in purely economic terms but also provide a powerful basis for the design criteria of efficiency, certainty and speed of technological developments. The intelligent machine, satisfying these criteria, sits comfortably at the centre of processes of production and control—a position deemed central for wealth creation. It is this position which is directly or indirectly used by government agencies, funding bodies and some AI researchers for their current preoccupation with machine-centred applications and research. Seen within this context, it is not surprising that decision makers express and justify human development in technical and economic terms.

This machine-centred definition of human progress may very well suit those AI researchers who are involved in the design of the intelligent machine and those powerful organizations who are involved in economic production and control. However, it gives no comfort or hope to those who are unemployed, disadvantaged, handicapped, poor or aged, for whom progress means meeting their daily human needs such as food, clothes, shelter and general human needs such as health, welfare, literacy, education, employment and training.

If we are seriously concerned with the applications of technology to human development and progress, then we should work towards transferring reasonable resources to meet the above human needs. This means that technological developments and AI research should consider these needs as their central focus. This focus is not only essential for human progress but it is also fundamental for sustaining and enhancing the social fabric of society.

If we are to design technologies which meet broad human needs, then they should aid human access to all relevant organizational structures of society and sources of knowledge maintained by those structures; enhance human life chances and opportunities; and also aid humans to participate in decision-making processes. For this to happen at the ground level, what is needed are those technologies which enhance human social communication and life skills and human cognitive competences. In other words, there is a need to provide education and training environments which enable such competences to be achieved by those groups in society who, at this moment, are neither participants nor direct beneficiaries of new technology, but are on the receiving end. Without this emancipatory use of technology, these groups are unlikely to attain expertise and skills suitable for participating in all aspects of society, sharing wealth, ensuring human dignity, equality of opportunity and justice and liberty. From this viewpoint, education programmes such as 'keyboard skills', computer literacy and computer-based training only tend to reinforce a machine-centred notion of production. It is worth noting Street's comments within the context of adult literacy which have a direct bearing on these issues:

> Production-related objectives are inseparable from other objectives concerned with transformation of economic and social structures. To define, 'functional' literacy according to these two categories, one economic, 'the rest' 'cultural', is to separate the 'inseparable' and, indeed, to define most of social life as a residual category of the economic.[12]

In this context, most of the AI work on expert systems and intelligent robots is only marginally relevant to emancipatory education. This is not to say that AI techniques such as user-modelling, intelligent tutoring systems[13] and episodically based knowledge systems[14] are not relevant to the design of socially beneficial education environments. However, the design of such environments requires not only appropriate AI techniques but also an understanding of the social and cultural contexts within which human learning takes place and is made meaningful. Learning is based on conventions of society and conventions derive their meaning from social and cultural contexts, knowledge and skills are acquired most effectively when attached to one's own goals, expertise and life experiences. Technology and technical curricula are less important for acquiring human cognitive competences than is a significant, interactive, dynamic human–machine relationship. It is thus clear that the design of technologies for learning should not only consider the individual's needs, expertise and life experiences but should also take into account knowledge of the social context and of the nature of social interaction. Forgas's comments on the cognitive process and the nature of knowledge are very relevant in this case:

> Human knowledge is a social product. Not only is formal 'science' a social enterprise (Kuhn, 1962), but the very ideas, theories, representations and reasoning which govern mundane events are themselves the outcome of extensive social cognitive processes. Indeed, it is

possible to argue that all cognition has social origins (Mead, 1934), and in turn, social living is based on the consensual cognitive representations of individuals.[15]

Summarizing, knowledge-based systems which aim to put human needs at the centre of their developments should be so designed as to focus on the inextricable intertwinement of benefit and accessibility. Such a system should:

1. Build upon the individual's life experiences and expertise, and aim to provide those skills, expertise, and cognitive competences which enhance the individual's life chances, opportunities and access to institutional structures and resources of society.
2. Take into account the individual's social, cultural contexts and constraints within which 'the learning' could be meaningfully interpreted and validated.
3. Consider those domains such as health, welfare, literacy, education, employment, law and the arts which not only are central to social progress but provide the individual with enhanced access to institutional resources for survival and opportunities for participating in increasingly complex technological societies.

The central focus of such a system should be transfer of knowledge rather than transfer of technology; sharing of knowledge and resources rather than transfer of information and monetary resources; education rather than technical training.

This focus should provide:

1. A framework for AI researchers to investigate issues of knowledge acquisition, representation and interpretation within broader social and cultural contexts in collaboration with researchers from areas such as social psychology, developmental psychology, anthropology, cultural history and the visual arts. Such developments are necessary if we are to design technologies for human learning rather than machine learning.
2. A challenge to those AI researchers concerned about social issues to provide a focus for a human-centred approach to knowledge-based systems, an alternative to the machine-centred approach.

REFERENCES

1. Feigenbaum, E. A., and McCorduck, P. (1984). *The Fifth Generation*, Pan Books.
2. Cooley, M. (1980). *Architect or Bee?*, Langley Technical Services, Slough.
3. Rosenbrock, H. (1982). *Technology Politics and Options*, EEC First Conference on Information Technology.
4. Noble, D. F. (1983). 'Democracy'. Report.

5. Galjaard, J. H. (1982). *A Technology Based Nation*, Interuniversity Institute of Management, Delft, The Netherlands.
6. Gill, K. S. (1984). 'Crisis and creation-computers and the human future'. In *Artificial Intelligence: Human Effects* (Eds M. Yazdani and A. Narayanan), Ellis Horwood.
7. Michie, D., and Johnston, R. (1984). *The Creative Computer*, Penguin Books.
8. SEAKE Centre (1984). 'Introduction'. The SEAKE Centre, Brighton Polytechnic.
9. Bruner, J. (1983). 'The growth of cognitive psychology: developmental psychology'. In *States of Mind* (Ed. J. Miller), British Broadcasting Corporation.
10. Simon, H. A. (1982). *The Sciences of the Artificial*, 2nd ed., The MIT Press.
11. Scott, D. (1985). *Polanyi, Polymath, Everyman Revised: The Commonsense of Michael Polanyi*, The Book Guild.
12. Street, B. V. (1984). *Literacy in Theory and Practice*, Cambridge University Press.
13. O'Shea, T., and Self, J. (1983). *Learning and Teaching with Computers*, The Harvester Press.
14. Schank, R. C. (1982). *Dynamic Memory*, Cambridge University Press.
15. Forgas, J. P. (1981). 'What is social about social cognition?'. In *Social Cognition* (Ed. J. P. Forgas), Academic Press.

Artificial Intelligence for Society
Edited by K. S. Gill
© 1986 John Wiley & Sons Ltd

3. A WAY FORWARD FOR ADVANCED INFORMATION TECHNOLOGY: SHI—A STRATEGIC HEALTH INITIATIVE

RICHARD ENNALS Department of Computing, Imperial College of
Science and Technology, London

ABSTRACT

As the Alvey Programme for advanced information technology moves
into its third year, the writer suggests one way forward after 1988, a
Strategic Health Initiative (SHI). After an analysis of current collaboration in the applications of artificial intelligence, an overview is
given of the potential in the field of health, and of current work with
expert systems. A number of research questions are raised for the
SHI, including implications for the patient and for social provision.
Practical suggestions are made for initial moves, and the issue is
placed in an economic and moral context in addition to technical
considerations.

THE CONTEXT

UK current strength in advanced computing research

The United Kingdom has a considerable reputation internationally in the field of advanced computing and, in particular, in the area known as 'intelligent knowledge-based systems'. This strength lies largely in the area of academic research, as British companies have not to date led in exploitation. Funding for this research has been erratic. British companies have invested little in research, development and training by comparison with their overseas competitors, in particular the United States and Japan.

British government research expenditure has been heavily weighted towards military applications. Fundamental research in artificial intelligence has rarely been in favour, and the 1973 Lighthill Report to the Science Research Council led to the withdrawal of a large proportion of the support for current projects and departure overseas of many outstanding research leaders.

The Alvey Programme of research in advanced information technology is a brave attempt to recover, with an innovative basis of collaboration between government, industry and academics. An extensive set of projects is now in place, developing the enabling technologies of intelligent knowledge-based systems, man–machine interface, software engineering and very large scale integration, as well as demonstrating their practical application. In particular, in the area of intelligent knowledge-based systems (IKBS), funds are now fully committed for the Alvey Programme up to 1988.

Thoughts have begun to turn to the issue of what happens next. Is there to be continued research after Alvey? What will be the emphases? Which application areas will be given special emphasis? Who will do the work? Who will pay?

UK National Health Service under threat

One significant area for the application of advanced technology is the Health Service. At the time of its establishment in 1948, the National Health Service was a model for international health care provision and a central part of the policy of a government concerned to strengthen its people after suffering and war, seeing such provision as an essential investment. It has since suffered from government neglect, with funding failing to match needs and hospitals not being equipped with the same level of technology that is standard in other advanced countries. It has become regarded all too often as an optional expense, increasingly to be devolved to the individual or the 'community', where the financial resources required for work with advanced technology are not available.

Collaboration in the current Alvey Research Programme

The present Alvey Programme involves a degree of collaboration which has no British peacetime precedent. Within government three separate ministries (Trade and Industry, Education and Science, and Defence) are learning to coexist, progressively resolving inconsistencies between their bureaucratic methods. Companies were unfamiliar with 'pre-competitive collaboration', and the last two years have been very much of a learning process. Many universities are participating in their first collaborative research with industry, as opposed to contract research. There have been exchanges and secondments of personnel, and new working relationships have been forged. Through a variety of schemes understanding has increased between industrialists and academics. In the strategic projects real collaboration has meant parity of esteem and mutual respect between the sides. The Alvey Programme has had to pioneer new forms of collaboration that could form the basis of research and development in other scientific fields, going beyond conventional funding through government research councils and commercial contract research.

Broadening Government Involvement

There is considerable potential for the involvement of further government departments. The Manpower Services Commission, part of the Department of Employment, is concerned with training of the young unemployed and with increasing the quality of training provision for clients in industry of all ages. The same technology that is being developed in the Alvey Programme can be deployed in training. For example, an expert system that knows about maintaining robots can be adapted to provide training in robot maintenance. A system that knows about regulations or legislation can explain them to trainee civil servants. There is considerable scope in the Department of Energy for the adoption of expert systems for oil and mineral exploration, whose potential has been demonstrated in the United States with PROSPECTOR and Dipmeter-Advisor. The Department of Education and Science could make considerable use of intelligent knowledge-based systems in education, building on work using micro-PROLOG and LOGO, and on broader work in intelligent computer-assisted instruction and expert tutorial systems. The morass of university entrance procedures and careers options could be made more penetrable with intelligent advice systems.

Perhaps the Department of Health and Social Services is the most likely candidate for involvement in research and development. Within the current Alvey Programme there are a number of relevant projects on which to build; for example:

1. A large demonstrator project led by ICL with the Department of Health and Social Services, Logica, Imperial College, Universities of Surrey and Lancaster. This is concerned with providing intelligent

decision support for DHSS officers and claimants, who are faced with complex rules and regulations which they have to apply to their own circumstances.

2. A large knowledge-based project in molecular biology led by the Imperial Cancer Research Fund, GEC and Brunel University. Work in cancer research has resulted in a significant body of knowledge of molecular biology, which needs to be made available to the doctor or researcher in response to a wide variety of possible questions.

What happens next?

The Alvey Programme does not end its current phase until 1988, but discussions and planning for what is to follow have already begun. Those within the Alvey Programme hope the collaborative process will continue, building on the successes and learning from the setbacks of this, the first British programme of its kind, in what is known as Alvey 2. Those who look for certain demonstrations of the efficacy of work in artificial intelligence after the first two years of a five-year initial programme may fail to find them, and advocate the redeployment of scarce resources to a different problem area, in a manner reminiscent of Sir James Lighthill's report in 1973 to the Science and Engineering Research Council, with an outcome known as Lighthill 2. More discriminating commentators may accept the power of the technology, but may wish to have activity less directed by research ideas, favouring 'applications pull' as well as 'technology push'.

In the United Kingdom the applications which pull are all too often military, with the subsequent civilian spin-off being ill defined. Defence research is justified, by its proponents, in terms of the scientific benefits of its wider application. This paper seeks to maintain that a stronger case can be made for the field of health care. Arguments have been presented elsewhere for a greater emphasis on applications in education and training. In both cases the argument is for investment in human resources, with a justification that is not merely expressed in economically quantified terms, but which has a strong moral, social and economic rationale.

Many leading researchers in artificial intelligence in the United Kingdom are not prepared to engage in defence-related work, and have been prepared to state their position in public. In the United States a large proportion of AI research is defence funded, and this position seems unlikely to change. The new American Strategic Defence Initiative is intended to provide financial support for a broad area of basic scientific research, further increasing the percentage of researchers with military support. The invitation is at present being extended to European companies and researchers. With the drying up of funds from the Alvey Programme in Britain, there will be undoubted pressure to participate in 'Star Wars' and perhaps a diminishing number of organized alternatives. Among those alternatives may be the European EUREKA initiative, exploring non-military applications of advanced technology with an emphasis on the needs of the civilian market, in collaboration with the more advanced members of the European Economic Community.

Researchers prefer to work on projects they believe in. Their brains cannot simply be hired for whatever purpose. At present skilled researchers are in short supply: they are well known for being able to command astronomical salaries overseas and for being transferred between research centres like football stars. Their choice of where to work need not be determined by money: after years of neglect and maltreatment they are suddenly in a new position of power where they can refuse work they find ethically unacceptable. They can choose instead to focus fundamental research effort on attempting to solve human problems.

In this spirit there follows a suggestion of a new initiative to tap this supply of idealism. We need a strategic focus for the next stage of development of an infant generation of technology, to the benefit of society in general—a 'Strategic Health Initiative'.

AN OVERVIEW OF THE STRATEGIC HEALTH INITIATIVE (SHI)

UK need to catch up in health defences

The United Kingdom has been falling behind. While our competitors have been raising their defences against illness and poverty, in Britain illnesses that had previously been eradicated are making their presence felt once more as a growing 'Fifth Column'. The conscription process for the First World War showed up the decrepitude of a large proportion of the population, meaning that they were unfit to fight. Many of the subsequent precautionary measures such as school milk and balanced school meals have been abandoned on short-term cost grounds, and the present population is physically becoming neither leaner nor fitter despite economic exhortations. Unhealthy foods are being marketed to the many for the commercial interests of the few. Officially commissioned scientific reports on the nation's diet are withheld from publication to avoid offending vested interests. National programmes of vaccination and preventive medicine are given little emphasis: prevention may be better than cure but the system of financial incentives is biased towards cure.

Where known enemy diseases threaten, our detection equipment is out of order. Straightforward tests are available for many forms of cancer, yet general scanning is not carried out on grounds of cost, and where intelligence of invasive disease has been acquired, all too often it is not transmitted to the individual concerned. The computer systems capable of managing the information exist but the funds are not provided to pay for them. We have the necessary technology for much of this work but have lacked the political will to apply it. To quote Ian Lloyd MP, 'We have found the enemy, and he is us.'

Our front line medical troops are pitifully resourced and are made to work inordinate hours in the medical trenches with substandard weapons. Patients have to be turned away from high-technology treatment in the cause of economy. Casualty wards servicing the M1 motorway are closed through lack of funds. Intensive care facilities are kept in mothballs. With changes in cleaning and catering arrangements, hospitals may not be healthy places to be if you are ill.

The non-commissioned medical ranks need expert advice, as they are all too often left in charge of a 'MASH' unit, providing intensive care without intensive training. They need access to the best technology where the medical expert is unavailable, and such technology needs to form part of their training.

Patients would rather not be ill and, if ill, would rather not trouble the doctor. Civil defence advice is needed for the patient in his home, aid in diagnosing the source of attacks of headache or nausea, preventitive measures to enable him to take evasive action, getting out of the line of fire of heart disease, cancer, or cirrhosis of the liver. 'Protect and survive' should be the watchword for the citizen in the blasted wasteland of community medicine.

Better coordination of resources

Often we have the resources available to repel an attack from outside, but they are not sufficiently organized. Doctors need decision support as they seek to define a strategy with a particular patient, and crisis management tools as numerous complaints emerge or as competing demands are made for finite resources. Increasingly they need a mastery of the official rules and regulations (on, for example, the prescription of certain drugs and their generic substitutes) and an encyclopaedic knowledge of drugs and their interactions. They need to be able to explain their diagnoses and treatment in appropriate language, based on a model of the level of knowledge of the patient and his family, and to draw on the experience of others. In the community medicine field, whether of barefoot doctors or a team of mobile professionals, information needs to be assembled, available and explicable. Advanced medical teamwork requires advanced information technology if the varied knowledge of the interdisciplinary team is to be brought to bear on shared problems.

Aids to independent health

Medical research has developed numerous aids for the disabled, some of which have not been made widely available for economic reasons. Many disabled people have been enabled to lead a normal life, including employment, with the aid of some prosthetic device. Artificial limbs and specially adapted keyboard input devices are well known, enabling people to make use of any controlled movement. Life for the blind or deaf, or even the deaf blind, is made more possible by language and communication systems. A current mathematics student at Imperial College is both deaf and blind and works with the aid of a computer with braille input and output, also using electronic mail.

With the advent of artificial intelligence techniques, further advances are made possible. Artificial intelligence is concerned with the study of human thinking and its modelling in computer programs. We can learn about particular problems by attempting to model them, and the consequent programs can be of use in helping people to solve such problems themselves. Early work has been done in psychiatry and psychotherapy, and in problems of vision and speech, which shows the potential for further work. Military funding has gone into systems for voice and speech recognition and for message understanding. An application focus in the field of intensive medical care or care of the multiply handicapped could be extremely beneficial, using, for example, speech-driven workstations as are being developed on an Alvey large demonstrator project.

ARTIFICIAL INTELLIGENCE IN MEDICINE TODAY

American medical expert systems

For twenty years research in the United States, where the majority of work in artificial intelligence to date has been done, has been undertaken into medical 'expert systems', systems embodying knowledge about a particular specialist aspect of medicine. The names of some of the best known are given below, together with their area of specialist application:

MYCIN	bacterial infections
CASNET	glaucoma
INTERNIST	internal medicine
VM	intensive care
PUFF	respiratory conditions
ONCOCIN	cancer

These systems have each taken many man-years to produce and have relied on access to highly expensive computer hardware. In recent years the cost and size of the necessary hardware has fallen dramatically and advances in software technology have made it much easier to develop systems for new specialist areas. Techniques have become more modular and transferable, and advances have been made in eliciting the knowledge of the expert which forms the basis of the system.

Artificial Intelligence and medicine in the United Kingdom

In the United Kingdom, partly for reasons of economic necessity, there has been an emphasis on what can be achieved with affordable hardware, regarding systems that are developed as tools to aid the practising doctor or nurse rather than as any kind of replacement. Dr John Fox of the Imperial Cancer Research Fund has developed an expert system shell called PROPS, which supports aids for the diagnosis of a number of cancers and related ailments, as well as being used in education and training. Current demonstration systems deal with ischaemic heart disease and cystitis. Dr Peter Hammond and Marek Sergot of Imperial College have extended PROLOG (the language used to write PROPS) to provide a flexible system called APES in which expert systems can be built. APES is one of the components of the diabetes management system under development by the London New Technology Network. These systems offer explanations of their diagnoses in terms of the facts and rules with which they had been provided and are available for commonly used personal computers. It is envisaged that within two or three years general practitioners and hospital doctors should all be offered the use of such systems.

RESEARCH QUESTIONS TO BE TACKLED ON THE SHI

Can a system be comprehensible, affordable and useful?

Given that the potential applications of medical expert systems have already been demonstrated, it is possible to develop systems that are at the same time comprehensible, affordable and useful? Can the same systems be useful for medical use and for medical education? (This had happened for MYCIN.) Can the same system be comprehensible to doctors with different approaches, and to the patient? What kinds of explanation are medically useful? How many doctors have chosen to use such systems? Do patients sometimes prefer their original consultation to be with a computer? (Weizenbaum found that his secretary preferred to talk to his ELIZA system than to him.)

How much involvement can the patient have in his own treatment?

Does the technology offer opportunities to improve the relationship between doctor and patient? Can the patient take the initiative and take a more active role in both diagnosis and treatment? Is there some information to which the patient should not have access? Can the system be made sufficiently 'friendly' for use by the patient? Is medicine less effective if it loses its mystique? We have evolved the heuristic 'principle of symmetry' in collaborative problem solving with the computer. Can there be a symmetry between the doctor and the patient (with the computer potentially standing in for either)? Do concepts of object-level and meta-level knowledge cast light on this problem? To what extent do we know how to represent the knowledge of the medical expert?

What are the implications of computer access for the disabled?

Can we revise our concepts of disability when new forms of communication and functionality are opened up? Can we provide voice input, touch-sensitive screens and hand-held devices like 'mice' to satisfy all needs?

Would we be offering an outlet for abilities that have not previously had a mode of expression? The experience with the use of Bliss symbolics to aid the communication of non-talking handicapped students and early experience with word processors for the disabled suggests that considerable abilities have been wasted. Are there implications for the provision of housing and training for the disabled? How will governments and officials be affected when the disabled are enabled to answer back?

What is the appropriate means of offering computer access?

Should computer access, like telephone facilities, be provided as a facility through local authorities? In France an attempt is being made to provide computerized enquiry facilities for all telephone subscribers. Could something similar be provided for the disabled and housebound? Should there be some kind of 'prescription system' for computer consultations? What precautions can be taken against the inappropriate use of computers in medicine? Will we have computerized medical malpractice cases?

What can current robotics offer the disabled?

Artificial intelligence offers the prospect of considerable enhancements to current prosthetic limbs, which can take on progressively more programmed tasks. Current commercially available domestic robots suggest that a similar process will happen to the revolution in personal computers, as prices drop with rising sales and simpler design. Will this remain subject to the private market or can we expect robots on prescription? We should take full advantage of increasingly intelligent machinery, but many aspects of care for the sick and disabled are more a matter of human contact, communication and response. In that sense the 'caring professions' should not be threatened by the new technology, but should find themselves more concerned with the human needs of the patient.

Are disease levels lowered by effective computer-aided screening?

The evidence overseas suggests that computer-aided screening can be highly effective, while equivalent work remains to be done in the United Kingdom. On a broader scale, epidemiology aided by medical signal processing has already brought results, predicting the geographical pattern of movement of diseases such as rabies and identifying the source of outbreaks of cholera. Work is beginning in the application of artificial intelligence techniques in this area. The implications are considerable for the relief of suffering (and consequent budgetary savings) both in the United Kingdom and, in particular, in Third World countries.

What are the implications of medical expert systems with hypothetical reasoning for preventative medicine?

If a patient can be confronted with the choices that face him/her at a given stage in an illness, this may affect his/her later behaviour and prognosis. An expert system may offer a richer and more acceptable environment for such issues to be explored. It may also help produce an improved relationship between doctor and patient. Current disputes about the association of tobacco smoking and lung cancer, or of excess fat consumption and heart disease, would be advanced if the evidence were more open to scrutiny and explanation. The fear associated with the possible side-effects of vaccinations could be alleviated by a system that could assign and explain the weightings of different factors. It is possible that many prescriptions for tranquillizers and antidepressants could be rendered unnecessary given the means of exploring the problems faced by the individual and the choices open to him. In this sense the expert system could serve as an extension of medical counselling in the hands of an experienced counsellor.

What systems should be standardly provided for the GP?

The General Practitioner cannot be expected to have a complete knowledge of all specialisms, but needs to identify signs and symptoms and to know how to proceed. Is it possible to develop an affordable system that would actually be used? Should such a system be standard, or would doctors be better advised to make a free choice? Early experience with PROPS suggests that doctors may find such systems to be of considerable practical benefit. We need extensive pilot studies.

How can medical education be enhanced by expert systems?

There has been some relevant experience in medical schools in the United States and Japan, and development work is clearly required, as in other areas of education and training. In the case of MYCIN, an expert diagnostic system, with the addition of a tutorial component GUIDON, is used for medical education. Doctors can gain vicarious experience in diagnostics without using real patients, with monitoring assistance and advice from the computer. Increasingly advanced technology systems are appearing in hospitals; with work in artificial intelligence we should expect the systems to be made comprehensible to the user, possibly through the addition of an 'intelligent front end'. This intelligent front-end program should incorporate expert knowledge of the subject specialist and should offer tutorial explanations of its working.

What can be done in occupational health?

Advisory systems could be envisaged for a number of occupational contexts, which could reduce accidents and illnesses related to working conditions. Considerable economic savings could be made through the strategic location of low-cost personal computers, which could both provide advice and monitor the information provided by workers. Fire, safety and building regulations are obvious candidates for representation as programs, as are official standard procedures. Professional bodies, trade unions and employers' organizations could all see the value of supporting such developments, and government should gain through paying less sick pay.

LAUNCHING A STRATEGIC HEALTH INITIATIVE

A Strategic Health Initiative would seek to draw on and advance progress in medical science, advanced computing and social administration. Its success would have enormous potential benefits, not only for the health of the nation but also for the economy, through the export of software, hardware and medical technology and know-how developed on such a national programme. Improved health and medical services would provide considerable financial benefits, as would the development of a better trained workforce. It would have direct effects on the whole population, bringing them into contact with computer technology in a benevolent context, reducing the division into two nations of 'haves' and 'have nots' with respect to health and computer literacy.

To start the programme would require the kind of emphasis on collaboration that we have seen in wartime, and which has been developing in the Alvey Programme. It would require the political will to identify the priority and to allocate the level of resources required. Alvey is costing a total of 350 million pounds over five years (200 million from government). This is of course as nothing against defence budgets, from which less tangible gains for the lives of civilians emerge. If we lack that will, then we become 'the enemy within', the 'cancer in the body politick'.

Richard Titmuss analysed societies in terms of how blood was transmitted in his book *The Gift Relationship*. In the National Health Service blood is donated without charge. A similar analysis could be given in terms of the transmission of knowledge and lies behind the campaign for a strong system of state education, free at the point of need. The proposal for a Strategic Health Initiative concerns the amalgamation of the two. The health of individuals is seen as integral to the health of the nation. Illness is not a crime to be punished by financial penalties, and information concerning the restoration of health should be freely available in accessible terms.

A Strategic Health Initiative would need to start on a pilot basis before it could expect to be adopted as government policy. The Alvey Programme teaches us, by its own omissions, that it is worth planning beforehand and establishing the groundrules for collaboration. Many existing voluntary bodies might see fit to collaborate in activity in the field, and organizations such as the Imperial Cancer Research Fund and the London New Technology Network have experience in uniting academics with medical and industrial needs. We might expect to elicit a response in terms of practical involvement from Community Health Councils, Citizens'

Advice Bureaux, single-illness charities such as those concerned with multiple sclerosis, heart and lung disease, cancer, leukaemia, medical pressure groups such as MIND and MENCAP, and the medical royal colleges and professional associations and unions. Paramedical professions such as speech therapists, occupational therapists and physiotherapists would be essential allies, and are already commencing their involvement with computer technology on a somewhat *ad hoc* basis. Funding should be attainable for a pilot phase through charitable foundations and pharmaceutical companies, as well as from the Medical Research Council, Science and Engineering Research Council, Economic and Social Research Council, Manpower Services Commission and Department of Health and Social Services.

If such an application of advanced computing technology to health is to take place, initial moves need to be made without delay. A committee should be established, drawn from some of the bodies above and from the various professional associations in the Health Service, including appropriate representation from industry. It should be asked to produce a report within a period of months, outlining a programme of action. Pilot studies should commence immediately in the relevant practical areas in advance of general government funding, and a nucleus organization should be established on the premises of a contributing body.

If such a programme were successful, the strategic results for the country could be spectacular. We could expect an improvement in the health of the population, with a cost-effective change of emphasis to prevention rather than cure, and a fall in the number of working days lost each year through illness. Industry could benefit from export sales of the resulting systems and the applications that followed in other sectors. The research community could benefit from the motivation of a continuation of work in 'advanced technology with a human face'.

Intelligent computer technology places a new burden on us to determine the kind of society in which we choose to live. It assumes the form laid down by its masters. If we abdicate from participation in the decision as to how the technology is to be used, we must accept responsibility for what follows. I close with the words of Lord Beveridge, whose work laid the foundations of the British Welfare State, including the National Health Service:

> The object of government in peace and in war is not the glory of rulers or of races, but the happiness of the common man.

Artificial Intelligence for Society
Edited by K. S. Gill
© 1986 John Wiley & Sons Ltd

4.

WILL AI LEAD TO A SUPER SOFTWARE CRISIS?

DEREK PARTRIDGE Computing Research Laboratory, New Mexico
State University, Las Cruces, New Mexico

ABSTRACT

Conventional software engineering has led to a software crisis: computer programs can produce results that are unexpected, incorrect and unexplainable. In other words, large computer programs, which are in control of ever increasing portions of society, can be both unreliable and incomprehensible. AI software extends the domain of problems addressed into that in which there is typically no complete specification of the problem, no clear-cut correct or incorrect answers, and a necessity for self-modification. In addition, the domain of AI problems includes many that hold much potential for societal disruption. AI software thus has all of the ingredients necessary for a super software crisis.

INTRODUCTION

I shall argue that the current techniques developed to mitigate the effects of the software crisis rest on certain assumptions that are inapplicable to much of AI. The resultant program development methodology is also not obviously applicable to AI. Incautious application of AI systems as practical software is, I argue, likely to aggravate the software crisis.

Attempts to structure AI program design to fit into, and thus benefit from, the conventional design methodology may be misguided. I argue for the development of a disciplined iterative program development methodology. I also point out the necessity for, and the dangers of, self-adaptive programs in AI.

SOFTWARE ENGINEERING AND THE SOFTWARE CRISIS

Since 1969 it has been widely recognized that the software industry has a major problem on its hands—the software crisis. The software crisis is a state of affairs in which we construct, sell and increasingly rely on computer programs that are incompletely understood. The large programs that are now in control of so much of our lives are both unreliable and not fully comprehensible.

The last decade or more has been, for software engineers, a period of searching for and developing techniques to alleviate if not eliminate the problems associated with practical software. Disciplined programming languages, such as Pascal, and the so-called 'structured programming' movement are just two products of the attempts to undermine the software crisis.

Another key idea is formalization of the program design methodology. There are many variants, but the underlying core can be caricaturized as the SPIV paradigm:

Specify → Prove → Implement → Verify

The 'correct' way to develop software is to first fully and formally specify the problem. This specification should then be checked for completeness and consistency (ideally it should be formally proved to be so). Only then should we design and code an implementation of this specification. Finally, the implementation should be verified against the specification.

That is the story; the practice tends to be much less formal. For example, implementations are tested rather than formally verified. Thus the implementations that we rely on are, at best, not correct, for free from known error. Here we have a major source of the software crisis, and it stems from our inability to adhere closely enough to a SPIV-based paradigm. The standard of excellence is clear and all effort is geared towards achieving it. Any departure from the SPIV exemplar is viewed as an unfortunate lapse that time and research may allow us to correct.

Figure 1 is a more complete illustration of the current practical methodology for software design. It is a far cry from the SPIV paradigm, but this is seen, as I have said, as a weakness that is to be corrected.

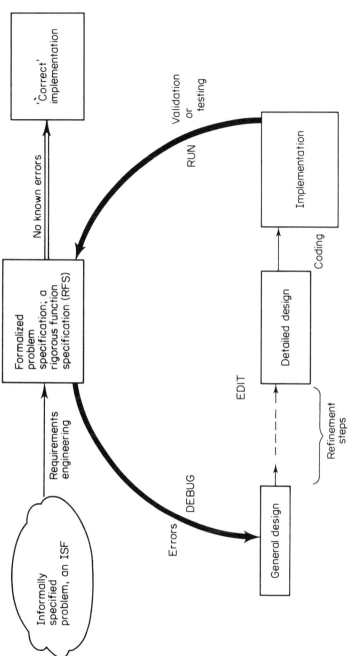

Fig. 1 The program development process of software engineering

From our current perspective there are two critical assumptions that are integral to this class of program development methodology:

1. The problem is completely specifiable (even if we cannot yet formally prove completeness and consistency).
2. The output from an implementation is decidably correct or incorrect.

These two key assumptions will be used to highlight the difference between AI software and conventional software.

AI SOFTWARE AND INCREMENTAL PROGRAM DESIGN

The paradigm at the heart of AI program development has no name, although terms like Run-Understand-Debug-Edit (RUDE) begin to capture the essential cycle involved. Rather than verifying an implementation of a complete specification, the procedure in AI can be better described as incrementally developing an adequate approximation to some incompletely specified function (ISF). One can agree with this coarse distinction but claim that its significance is not the exposure of an alternative paradigm for design; instead its significance is a reflection of the 'hacking syndrome' that is endemic in AI. When the field reaches maturity, a state whose advent will be hastened by the current surge of interest in commercial AI, the RUDE paradigm will be replaced by a SPIV-based methodology. AI system designers will then have become software engineers distinguished only by the complexity of their domain.

This view, I argue, is at best premature and at worst totally wrong: it may be that a RUDE-based methodology is an approach well suited to the nature of AI problems. That is not to say that typical manifestations of RUDE techniques are satisfactory—far from it. It is only to say that the general nature of the RUDE paradigm may be more appropriate as a basis for AI design than is the SPIV paradigm. The RUDE paradigm is in dire need of development, but that does not necessarily mean development towards a SPIV-based methodology.

Complex and ill-structured problems are the domain of AI, and the design of adequate implementations of such problems necessarily appears to be an incremental, evolutionary, exploratory process. Instead of striving for complete specifications and the verification of proposed implementations, we should concentrate more on incremental development of specifications as a result of assessment of performance. This in turn depends on development of abstraction techniques (in order to obtain intellectually manageable specifications from behaviourally interesting programs) and techniques of design for change.

Figure 2 is a schematic illustration of a RUDE-based program development methodology. Notice that now we no longer have a complete specification of the problem against which an implementation can, in some sense, be measured. We have an ISF and we are trying to discover an

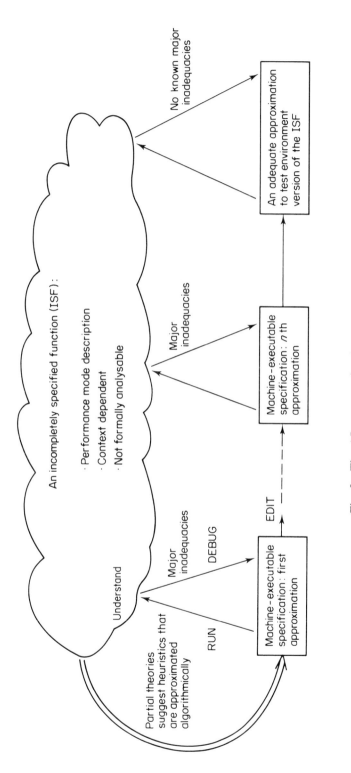

Fig. 2   The AI program development process

adequate approximation to it. So we have lost the first key prerequisite with which software engineers hope to overcome the software crisis.

The important point here is that we have not lost this key feature because AI programmers are incompetent and have failed to appreciate the fundamentals of software engineering. A complete and precise specification of an important class of AI problems is, I shall argue, just not possible, even in principle. Closely associated with the lack of a complete specification we also lose that other touchstone of software engineering: decidability of program output. In AI we must typically work with results that are, at best, adequate, rather than correct.

AI PROBLEMS AND CONTEXT SENSITIVITY

Many AI problems are highly context sensitive, and the relevant context is not easily circumscribed. Worse than that, many AI problems exhibit tightly coupled context sensitivity: intelligent answers to such problems are highly dependent on contextual information, i.e. information external to the particular problem at hand. Thus, for example, the meaning of a sentence may be minimally dependent on the actual words that constitute the sentence. The interpretation of an image may also have little dependence on the actual image data itself. The best explanation of the reasoning behind some decision can be more dependent on who wants the explanation, and why, than on the particular decision itself, and so on.

A succinct and complete context-free specification of such problems thus appears to be an unrealistic goal (although there is always hope that such apparently tightly coupled, context-sensitive problems can be decomposed into a collection of more manageable, fairly modular sub-problems).

An example of a non-tightly-coupled AI problem is chess: the next best move for any board configuration is (almost) totally dependent upon the particular configuration and independent of any contextual considerations.

This problem is now commonly seen in connection with the most successful application domain of AI—expert systems. The design of such systems must deal with incomplete knowledge bases (no one seriously suggests that a complete knowledge base is possible) and with incrementally updating these knowledge bases. They need logics for inferencing from incomplete knowledge (as proposed, for example, by Levesque[1]) and they need a paradigm for incrementally updating the knowledge without generating an unmanageable tangle—i.e. expert system designers seem to have a clear need for a RUDE-based paradigm.

The necessity for a RUDE-based methodology for AI system development does not bode well for efforts aimed at combating the software crisis. In addition, before we can hope to exploit anything like the full potential of AI in practical software, that software will have to be self-adaptive in a non-trivial sense.

SELF-ADAPTIVE SOFTWARE AND AN ESCALATION OF THE SOFTWARE CRISIS

Self-adaptive software translates into a need for machine learning in AI. A number of observers (for example, Samuel[2] and Schank[3]) see machine learning, despite years of neglect, as critical to the design of AI systems. I have already argued at length and in detail[4] for this and for most other points that I have had to deal with very cursorily here. Reasons why I believe that a non-trivial machine learning capability must be accommodated in a paradigm for the design of AI systems, and why such a belief implies the RUDE rather than the SPIV paradigm, are:

1. Everyone is different; if an AI system is to behave even reasonably intelligently, it cannot neglect this fact.
2. Any one person is different at different times; a major role of AI systems is to impart knowledge to people, and if it does not respond to the changes it induces, it will have failed—it will not be AI.
3. The empirical world is a rapidly and subtly changing place (quite apart from the people in it); and an AI system to remain an AI system must keep abreast of the relevant changes.

The discovery of robust and generalized learning algorithms would perhaps enable the necessary adaptivity to be accommodated within the SPIV paradigm. Until such time as we find these learning algorithms (and I do not think that many would argue that such algorithms will be available in the foreseeable future) we must face the prospect of systems that will need to be modified, in non-trivial ways, throughout their useful lives. Thus incremental development will be a constant feature of such software and if it is not fully automatic then it will be part of the human maintenance of the system. In addition, the societal problems that could be caused by imperfect machine learning (and that is what we will have) hardly bear thinking about; I have discussed this particular potential source of societal disruption elsewhere .[5]

Hewitt[6] argues for what I take to be some RUDE-like paradigm for 'developing the intelligent systems of the future'. He discusses the problems of continuous change and evolution and the need to accommodate necessarily incomplete information, which implies, he says, exploration rather than searching—the traditional approach in AI.

By way of contrast Mostow[7] argues against a RUDE-based methodology and offers an alternative scheme, called COURTEOUS, which involves automating the process of implementing a specification. I have pointed out both a misconception and a major problem with Mostow's suggestion.[8]

SUMMARY

The software crisis is due to our ability to construct and use programs of a size and complexity that are beyond the limits of human comprehensibility.

These programs are never guaranteed to function correctly, although with certain key assumptions abstract proof and verification may be possible in the future. In AI we lose even the prerequisites of proof and verification. The techniques developed in software engineering, such as structured programming, are predicated on a SPIV-based development methodology. AI appears to be best supported by an incremental program development methodology; such incremental development is not a useful basis for the all-at-once techniques that have been developed. So if incremental development is crucial in AI, as I believe it is, then the currently available tools for minimizing the impacts of the software crisis will not be of much use; we will need to develop analogues appropriate for an incremental paradigm.

Sandewall,[9] for example, has raised this last point with respect to structured programming and AI. The loose conglomerate of rules and guidelines known as structured programming is an all-at-once technique, which is typical of work within the SPIV paradigm. Given a complete specification of some problem, how can we implement it such that the resulting object is maximally transparent to humans? That is the question that structured programming seeks to answer. If any changes are introduced they are introduced into the specification, which is then reimplemented (at least that is the theory; the practice is all too often some manifestation of RUDE techniques of the worst kind).

By way of contrast, the RUDE paradigm suggests a need for some analogous but different concept—Sandewall calls it 'structured growth'. Such a concept would involve rules and guidelines for incrementally adding structured code in a way that maintains overall clarity of structure and for altering parts of a structured object, again in a way that preserves its perspicuity.

Added to these problems associated with the basic paradigm, we have a need for self-adaptive software. Self-modifying programs have nothing positive to add to the problems of understanding and predicting the behaviour of programs. All in all, unless we are very cautious in our attempts to implement AI software, the software crisis could well escalate to something much worse—a super software crisis.

Apart from a general caution and a thorough testing of, say, machine-learning algorithms, prior to practical application, I would suggest that in order to avoid a super software crisis, and yet eventually exploit the full potential of AI, we must develop a discipline of program design that is based on the RUDE paradigm. However cautiously we proceed with the production of AI software, we must also recognize that intelligent problem solvers (mechanical or organic) will not be foolproof; checks and balances will always be needed—perhaps all the more so with AI that has not been subjected to the selective pressures of a long evolutionary process.

REFERENCES

1. Levesque, H. J. (1984). 'The logic of incomplete knowledge bases'. In *On Conceptual Modelling* (Eds M. L. Brodie, J. Mylopoulos and J. W. Schmidt, Springer-Verlag, New York.
2. Samuel, A. L. (1983). 'AI, where has it been and where is it going?' *Proc. IJCAI* 1983, Karlsruhe, West Germany, pp. 1152–1157.
3. Schank, R. C. (1983). 'The current state of AI: one man's opinion'. *The AI Magazine*, Winter/Spring **1983**, 3–8.
4. Partridge, D. (1985). *Artificial Intelligence and the Future of Software Engineering*, Ellis Horwood, Chichester (forthcoming).
5. Partridge, D. (1985). 'The social implications of AI'. In *Artificial Intelligence: Principles, Applications, and Implications* (Ed. M. Yazdani), Chapman Hall (forthcoming).
6. Hewitt, C. (1985). 'The challenge of open systems'. *BYTE*, **April**, 223–242.
7. Mostow, J. (1985). 'Response to Derek Partridge'. *The AI Magazine*, **6**(3).
8. Partridge, D. (1985). 'AI is a behavioral science', *The AI Magazine* (forthcoming).
9. Sandewall, E. (1978). 'Programming in an interactive environment: the "LISP" experience'. *Computing Surveys*, **10**(1), 35–71.

PART 2

AI—Philosophical Issues

Artificial Intelligence for Society
Edited by K. S. Gill
© 1986 John Wiley & Sons Ltd

5. WHY AI CANNOT BE WRONG

AJIT NARAYANAN Department of Computer Science, University of Exeter

ABSTRACT

Asking why AI cannot be wrong is not the same as asking why AI must be right. Instead, the question is meant to focus attention on two widely held beliefs concerning AI research: that AI is a science and that AI is making progress. The first belief concerns the objectivity of AI, and it will be argued that AI does not satisfy traditional criteria for what it is for a discipline to be a science. The second belief concerns recent developments in AI and the increasingly common view that AI is making progress, especially in the area of expert systems. It will be argued that the criteria for judging whether AI is making progress assume much more than is justified. The conclusion will be that AI appears to be irrefutable, scientifically, and that progress in AI is impossible unless clearly defined objectives and goals are specified. Finally, the implications of this conclusion will be discussed and the dangers of 'blind' AI research highlighted.

AI RESEARCH

There is remarkably little concensus on the part of AI researchers as to what exactly it is that they do when they carry out AI research. The historically interesting, but now unpopular, view that AI '. . . is the science of making machines do things that would require intelligence if done by men'[1] was considered the least controversial[2] simply because it was the most empty. That is, it left open the crucial question as to which activities did require intelligence, and why. Here is a much more modern view: '[AI] is the study of mental faculties through the use of computational models'.[3] Note the use of the words 'science' and 'study' in the above quotations.

Some authors and commentators do not even bother to give their own views on what it is that they are doing when they carry out AI research. For instance, the editors O'Shea and Eisenstadt[4] assume that the reader already has knowledge of what AI is or can be introduced to what AI is by reading the papers in the book. The papers, by and large, describe the various programming languages ('tools') available, practices used ('techniques') and systems ('applications') implemented. That is, the book describes the technology rather than the theory of AI.

Others do not even consider that it is important to say exactly what AI is: 'I don't think it matters at all whether or not AI is a discipline or where its boundaries might be'.[5] Many commentators believe that AI is essentially multidisciplinary:

[AI] is inherently multi-disciplinary, and the more AI is applied the more disciplines will be involved. There is a central use of AI/Cognitive Science which is the systematic study of actual and possible intelligent systems.[6]

AI, broadly conceived, is just too large to be a single discipline—mainly because intelligent perception and behaviour touch so many aspects of computer science, control theory, and signal processing theory.[7]

Bobrow and Hayes, the editors of the responses, sum up the part of the questionnaire dealing with AI methodology as follows:

The distinction between on the one hand AI as a science, essentially part of Cognitive Science, with its focus being theories of intelligence; and on the other hand AI as a technology, part of Computer Science, with its focus being on the design of systems, ran through several responses.[8]

The aim of this paper, apart from trying to steer well clear of terminological issues, such as the distinction between 'science' and 'study', is to demonstrate that unless AI is provided with a proper theoretical basis and an appropriate methodology, one can say just about anything one wants to about intelligence and not be contradicted; unless AI is provided with some reasonable goals and objectives little of current AI research can be said to be progressing.

RECENT OPPOSING METHODOLOGIES

In 1966, Hempel[9] argued that scientific knowledge is arrived at by inventing hypotheses as tentative answers to a problem under study and then subjecting these hypotheses to empirical test. These hypotheses should be at least testable in principle if current technology cannot provide the tools for such testing. The confirmation of a hypothesis will normally be regarded as increasing with the number of favourable test findings. The hypotheses should not be saying something about phenomena that are in principle unobservable or unmeasurable, according to Hempel. However, Hempel also allows 'bridge principles' to link the hypotheses or theory with observable or measurable aspects of physical systems.

For instance, a theory in psychology may well use such indirectly observable theoretical constructus as 'ego', 'superego' and 'id': the purpose of bridge principles would then be to link these theoretical constructs with experimental observables indirectly, either by invoking previously established (i.e. tested) theories or by predicting what sort of physical behaviour would be manifest.

These predictions can then be tested and, if the predictions turn out not to be accurate, modified. There will usually be a limit to the amount of prediction modification in the face of adverse evidence. Scientists usually recognize when *ad hoc* modifications to a theory have reached a limit by realizing that the theory has become so complex that an alternative, simpler theory should be sought. Hempel also argues that a scientific theory deepens our understanding by predicting and explaining phenomena that were not known when the theory was first formulated.

Popper is another philosopher of science. He also argues that a scientific theory should be tested by empirical application of the conclusions that are deductively drawn from it. That is, predictions are formed which are easily testable or applicable.[10] If the results of practical experiments show the predictions to be verified, we say that the theory, for the time being, passed its test. However, if the predictions are falsified, then the theory from which they were drawn is also falsified. This gives rise to the notion of 'falsifiability': a scientific theory is not required to be capable of being singled out once and for all in a positive sense (verified); rather, its form should be such that it can be singled out. By means of empirical tests, in a negative sense. Thus, according to Popper, it must be possible for a scientific theory to be refuted by experience. If a theory is not falsifiable it is metaphysical, i.e. it is not saying anything about the physical world at all.

Hempel and Popper are usually regarded as being at two ends of a spectrum as far as their views are concerned. They disagree sharply on what the status of a theory should be whilst tests are taking place. Hempel's views are the more orthodox: a scientific theory is true if confirmatory evidence is found for it. Modification of the theory is allowed to take into account any hiccups in the evidence, although such modifications usually have a limit set by the increased and undesirable complexity of the theory. The task of the scientist is to confirm the theory. For Popper, on the other hand, a scientific theory is characterized by the possibility that a single negative piece of evidence will refute it. The taks of the scientist is to refute the theory, and the scientist should try any number of ways to falsify the theory. *Ad hoc* modifications of the theory are not allowed by Popper.

What is remarkable about these two opposing methodologies is the agreement they both share concerning the construction of a theory in the first place, and the requirement for a scientific theory to have some 'cash value' in the real world. That is, they both attack a view of science which may be called 'the narrow inductivist conception of scientific enquiry', which essentially holds that a scientist first collects data through observation and then generalizes from these observed facts to a theory. Hempel claims that scientific theories are not derived from observed facts but invented by the scientist to account for them. Popper believes that scientists do not wait passively for repetitions and regularities but instead actively impose regularities upon the world by inventing theories—conjectures—that were to be eliminated—refuted—if they clashed with observations.[11] Hence they both agree that the role of a scientist is much more active and inventive than the role assumed by the narrow inductivist approach.

They also both agree that the theory must have some applicability in the real world. Conclusions and predictions are derived from the theory and these must be subjected to empirical testing. The results of the tests will then have further implications for the theory.

WHERE DOES AI FIT INTO ALL OF THIS?

It seems to me that, given the above views, AI occupies a highly ambiguous position if it is claimed that it is a science, or a study, or mental faculties through the use of computational models. One popular view in AI is that AI programs embody a theory. For instance. Cendrowska and Bramer claim that 'A particular feature of AI programs is that they are generally written to illustrate a particular theory, and if judged successful it is this theory that will be featured in published accounts'.[12] Hayes also adopts a criterion based on program implementability:

> AI's criterion [or rigour] is not experimental corroboration, but implementability. An acceptable explanation of a piece of behaviour must be, in the last analysis, a program which can actually be implemented and run. And such an explanation is a good explanation just to the extent that the program, when run, does indeed exhibit the behaviour which was to be explained.[13]

It can be argued that the criterion of implementability is vacuous at the level of the Church–Turing thesis. There are many different versions of the thesis, but Hofstadter's version is the most appropriate here: 'Mental processes of any sort can be simulated by a computer program whose underlying language is of a power equal to [a language] in which all partial recursive functions can be programmed'.[14] This essentially states that any mental process which can be described by an algorithm can be executed on a computer. It is called a thesis because it cannot be proved to be the case; nevertheless, no evidence has arisen to the contrary. Hence, it may appear that as long as the AI researcher takes care to construct theories which can be written down as a sequence of algorithmic or computational steps, these theories can be implemented.

However, the vacuity of the implementability criterion becomes apparent when one realizes that in AI research the fundamental assumption is that computational concepts can be used to describe the workings of the human mind. Hence, an AI researcher will use the concepts of computational theory to construct AI theories. This leads to the conclusion that all AI theories will then necessarily be implementable on a computer, since computational concepts are used. It therefore follows that implementability *per se* leads to a self-perpetuating methodology: an AI theory is necessarily implementable, therefore implementability as a criterion is vacuous.

However, the authors of the above quotations have more in mind than this, which is that the results of the computer program should exhibit the behaviour or produce the results that were predicted of it. How were such predictions made?

One way to make predictions is for a human to work through the algorithm (theory) by hand and produce a set of results for a certain input. In this case all the implementation will show is whether the human's working through of the algorithm (theory) is correct. The use of a computer in this way may be useful for checking a human's ability to work logically through a set of instructions, but there would then be no justification for claiming that the results had an empirical validity over and above this simple checking exercise. The computer results are correct if they match the results of a human working through the same algorithm, and computer results can sometimes show that the human is wrong in his or her calculations. This only gives us an indication of how closely matched human and computer performance of the same algorithm are and says nothing about how accurate or relevant a description of human mental capability the theory (algorithm) itself is.

Another way to make predictions is for the human, once an AI theory has been constructed, to predict what he or she would like to see the program do. This method therefore does not rely on a human working through the algorithm (theory); rather, it allows an AI researcher to use the program as a tool for testing and sharpening his or her own theories.

One suspects that this is really the point the authors of the above quotations wish to make. However, what happens if the results produced by the program are not the expected or predicted results? It is a well-known fact in computer science that programs never work the first time. Just how much modification and fine-tuning of the program is allowed?

Hempel argued that *ad hoc* modifications to a theory were limited by the increased complexity of the theory and that after a certain threshold level of complexity was exceeded scientists would naturally and logically pursue simpler alternative theories. Popper argued that no *ad hoc* modifications of any sort should be allowed to a theory. If it did not work for one instance, the whole theory should be discarded. Popper's view may appear too strict for AI researchers (and computer scientists!), who would naturally believe that any computer program requires time to debug and test before being used for serious applications. Hempel's views could be of use here to those AI researchers who believe in the implementability criterion. Perhaps the main requirement for AI researchers is some criterion of program complexity which would allow them to judge just how complicated their programs are becoming as modifications, kludges and fine-tuning take place. Certainly, without such a criterion, there is a great danger that AI programs end up being highly complex and inconsistent in their theory as well as in their behaviour, simply because AI researchers continually modify their programs so that the desired and expected results are produced. If an AI program is or embodies a theory, and if AI is a science, then according to Hempel each modification will make the program/theory more complex, and eventually the program will be rejected in favour of a simpler theory.

In any case, even if a criterion of complexity for AI programs (theories) can be found, there still remains the suspicion that no criterion exists or can exist for determining whether an AI theory is true or accurate. It is probably this suspicion which has led many AI researchers to adopt the stance that their theories represent possible worlds rather than actual worlds—that an AI program represents a possible discription or explanation of mental faculties rather than an actual one. This sounds attractive until one begins to delve more deeply into the logic of possible worlds.

In modal logic, the following equivalences are valid:

'*x* is necessary' is equivalent to 'it is not possible for *x* to be false'
'*x* is possible' is equivalent to 'it is not necessarily the case that *x* is false'

These equivalences hold for individual propositions. From these basic equivalences we can build the notion of worlds. A proposition is necessarily true if it is true in every possible world. A proposition is possibly true if it is true in some possible world. Possible worlds are usually interpreted to mean conceivable worlds, states of affairs that can be envisaged. There are many different ways of looking at possible worlds, but one logical fact is clear: even within a possible world certain propositions can be said to be true or false. That is, even a possible world has certain fundamental principles concerning the truth and falsity of propositions, and these principles can be shared across possible worlds. If AI researchers are to talk of their theories representing possible worlds, they still need to have criteria of truth and falsity in their worlds. Until such time as AI researchers who claim that their theories represent possible worlds can provide such truth criteria which allow propositions in their own worlds to be true or false, the suspicion remains that the notion of possible worlds is being used to disguise the self-perpetuating and essentially irrefutable nature of AI research.

THE FUTURE FOR AI

In the sections above, the argument was presented as to why AI is not a science. Let us now turn our attention to the view that it does not really matter whether AI is a science or not: the important question concerns whether AI is producing the results. That is, many AI researchers would regard the above discussion as worthy of philosophers and navel contemplators in general. Instead, they may point to past and current work in AI as demonstrating that AI is progressing and is actually producing the desired results—science or no science.

There are very many AI projects now under way in universities and industries around the world, and new AI systems are appearing fairly regularly. Some of these systems are called AI systems because they use a few principles that, for historical reasons, are considered to be AI principles (e.g. alpha-beta pruning, search strategies). Others adopt an AI programming language, such as LISP or PROLOG and also use knowledge representation techniques, such as semantic networks, scripts and production rules. Although there may not be clear criteria for distinguishing AI programs from non-AI programs ('AI programs are those which embody AI techniques', for example), what criteria could we use for deciding that AI is making progress? If AI were a science, we could use such concepts as 'greater explanatory power', 'more generalized and simple', 'is consistent with a larger body of facts' and 'greater predictive power' to judge whether one AI theory is 'better' than another. If AI theories are replaced by better ones, we would have grounds for saying that AI is making progress.

Given my comments in the last section, it appears that there can be no scientific reasons for claiming that one AI theory is better than another and, hence, that AI is making progress, simply because the conceptual tools for measuring one theory against another, and so for measuring the progress of AI, are missing. Answers to such basic questions as 'What are the objective of AI?', 'What methods exist for achieving the objectives of AI?', 'What exactly is the relationship between an AI theory and a mental faculty?', 'Does the theory model, or describe, or explain the faculty?', 'How exactly can the model, or description, or explanation be evaluated?' and 'How will one know that the objectives of AI have been reached?'

are not forthcoming from AI researchers because, one suspects, they simply do not know. For an AI researcher to claim that such questions are unanswerable leads to the idea (which AI researchers would like to promulgate) that AI as a science will develop as results start to flow in and theories and methods are induced from these results.

This is precisely the notion that both Popper and Hempel agree is wrong. That is, they both agree that the narrow inductivist conception of scientific enquiry is wrong, but this conception is precisely that which is being adopted by those AI researchers who adopt a 'wait and see' attitude.

If AI is not a science, then how is progress in AI to be measured? By faster programs, speedier results, the use of parallel, distributed architectures? None of these would appear to give us grounds for claiming that AI is progressing, since they depend only on factors outside AI and not within, i.e. on hardware and software developments in computer engineering and computer science. Just because the technology on which AI depends improves does not necessarily mean that AI itself is improving, just as a racehorse improving its performance race after race does not necessarily mean that the jockey also improves. That is, it is the use to which the technology is put that can function as a criterion of improvement in AI.

Again, many AI researchers I have talked to are remarkably reticent on this point. Although they claim to be able to recognize improvement in AI there is very little clear expression of how this recognition is achieved. One view is that AI is throwing up some completely surprising results, as if the function of surprise is somehow part of the measuring tool. That is, it could be argued that the ability of an AI program to surprise its user with unexpected, but correct, results can be measured and somehow used to determine whether AI is progressing. This sort of tool would probably be most useful for deciding whether expert systems are progressing.

Let me relate a little story here. One of my PhD students wrote a natural language parser for a subset of the Arabic language. He used Prolog as the programming tool and definite clause grammars.[15] I looked over his shoulder as he demonstrated his system to me for the first time. The first two or three Arabic sentences were nicely parsed uniquely and the results, which consisted of categorizing each word in the Arabic sentence into its syntactic role (e.g. determiner, noun, and so on), as well as gender (male, female) and number (singular, dual, plural), were exactly those expected. Then one particular Arabic sentence was input. Whilst we waited for the system to respond (about 20 seconds), my student confidently predicted what would happen. Sure enough, the result of the parse was consistent with his prediction. Each time our version of Prolog finds a result and displays it, it then waits for the semicolon key to be pressed as a continue key so that it can find more answers to a query (very useful for database retrieval work). My student pressed the continue key, confidently expecting there to be no more answers. He was surprised to find the system returning exactly the same answer. He looked at it and then said: 'Yes, of course, I'd forgotten that this sentence can be parsed in two ways, depending or whether you believe the subject or object noun comes first.' He pressed the continue key again, and to his surprise a third parse was found. He looked at it closely, and then he said: 'Yes, the sentence has a third reading, depending on which interpretion you put on the verb. I hadn't realized the system could do that.' He pressed continue and again the same line was displayed. By now, he was becoming as sceptical as I had been since the first duplication. Whilst he was using all his Arabic knowledge to provide a fourth reading. I leant forward and pressed continue a few times. Each time, the same line was displayed. It was now obvious that the system was in a loop and that the same result was being displayed over and over again because of this loop and not because of backtracking.

The point of the above story is this: surprise is a very dangerous criterion to use for measuring AI progress as long as an appropriate AI methodology is lacking. Without such a methodology, it may be very difficult, if not impossible, to separate genuine surprisability from programming error fuelling user enthusiasm. The minimum step must be proving AI programs 'correct', in some sense of the word, so that surprising results cannot be caused by a bug in the program. But again, few AI researchers seem to be interested in the highly important area of AI software engineering.

CONCLUSIONS

What is all this leading to? I do not believe that it is necessarily a bad thing if AI is not a science. If we consider AI to be a technology rather than a science, many of the more mind-expanding claims for AI can be discarded. AI technology is the use of AI techniques and tools in real-world applications. AI techniques and tools are techniques and tools proposed by researchers who believe that they are providing the means whereby computers can be made to behave (not think) like humans. This constrained view of AI, which may well be attractive to those researchers who get their hands dirty, has some severe implications for the AI academic community.

First, there may be nothing in common between one AI project and another. For example, two different expert systems may have absolutely nothing in common except a name. No underlying principles of knowledge representation and inference need be shared between the two expert systems. No attempt need be made to relate expert systems with human experts. Expert systems are judged individually on the results they produce, and these results need not be judged against human expert's results—and similarly with natural language parsers, computer-aided instruction packages, vision systems, and so on. A system will survive if it finds a niche in the market-place; otherwise it is soon forgotten. No attempt is made to relate AI with explaining real-world phenomena and no attempt is made to look for underlying scientific principles.

Second, and much more importantly, if AI is mere technology, any attempt to pursue AI research for the sake of science will not succeed. Much has been made of the dangers of AI to mankind. If AI is a science, AI researchers can ignore these dangers by claiming that science had to be pursued for science's sake, in the same way that scientists working on nuclear fission argued that science had to be pursued. However, if AI is a technology, it can be held accountable and can therefore be constrained. Few people, even those deep in the Central Electricity Generating Board, would argue that nuclear reactors had to be built for the sake of building nuclear reactors because nuclear reactors are a technology. Moreover, they are a technology that evolved out of the science of nuclear and particle physics. Before a nuclear reactor can be built, a lengthy process is entered into whereby the CEGB needs to argue its case before the public. Local inhabitants and environmentalists can spell out the dangers of the new technology, and the chairman of the enquiry decides.

Perhaps this model would do nicely for AI technology. Before a company can launch an AI project which could have severe implications for employment and human self-esteem, it has to submit its proposal for vetting. The public can raise objections and an independent chairman decides. On no account is the argument to be allowed that the project should go ahead because AI is a science and so needs pursuing at all costs.

Finally, there is one big advantage to AI being a technology rather than a science: it can fail or succeed depending on how many people but it. Here lies the ultimate safeguard. No one can do much about electricity that is generated by nuclear reactors, because the GEGB holds the monopoly on electrical power in this country, but we can decide whether we want to buy AI goods or not, and what sort, as long as AI technology is not monopolized by central government. Hence the final power to decide what to do with AI rests in the best hands: ourselves.

REFERENCES

1. Minsky, M. L. (1968). *Semantic Information Processing*, MIT Press, p. v.
2. Boden, M. (1977). *Artificial Intelligence and Natural Man*, Harvester Press, p. 4.
3. Charniak, E., and McDermott, D. (1985). *Introduction to Artificial Intelligence*, Addison Wesley, p. 6.
4. O'Shea, T., and Eisenstadt, M. (Eds) (1984). *Artificial Intelligence: Tools, Techniques and Applications*, Harper and Row.
5. Feldman, J. (1985). In response to question 5 in a questionnaire distributed by D. G. Bobrow and P. J. Hayes (Eds), *Artificial Intelligence*, **25**, 376.
6. Sloman, A. (1985). In response to question 5 in a questionnaire distributed by D. G. Bobrow and P. J. Hayes (Eds), *Artificial Intelligence*, **25**, 377.
7. Nilson, N. (1985). In response to question 5 in a questionnaire distributed by D. G. Bobrow and P. J. Hayes, *Artificial Intelligence*, **25**, 376.
8. Bobrow, D. G., and Hayes, P. J. (Eds) (1985). *Artificial* Intelligence, **25**, 380.
9. Hempel, C. (1966). *Philosophy of Natural Science*, Prentice Hall.

10. Popper, K. R. (1972). *Objective Knowledge*, Clarendon Press.
11. Popper, K. R. (1959). *Conjectures and Refutations*, Routledge.
12. Cendrowska, J., and Bramer, M. (1984). 'Inside an expert system: a rational reconstruction of the MYCIN consultation system'. In *Artificial Intelligence: Tools, Techniques and Applications* (Eds T. O'Shea and M. Eisenstadt), Harper and Row, p. 454.
13. Hayes, P. J. (1984). 'On the differences between psychology'. In *Artificial Intelligence: Human Effects* (Eds M. Yazdani and A. Narayanan), Ellis Horwood, p. 158.
14. Hofstadter, D. R. (1979). *Godel, Escher, Bach: An Eternal Golden Braid*, Harvester Press, p. 578.
15. Pereira, C. N., and Warren, H. D. (1980). 'Definite clause grammars for language analysis—a survey for the formalism and comparison with augmented transition networks'. *Artificial Intelligence*, **13**, 231–278.

Artificial Intelligence for Society
Edited by K. S. Gill
© 1986 John Wiley & Sons Ltd

6. ETHICS, MIND AND ARTIFICE

STEVE TORRANCE Cognitive Studies Programme, University of
Sussex, Brighton

ABSTRACT

This chapter explores the difference between two kinds of claim that
can be made on behalf of AI: *the narrow claim*, that certain specific
mental processes—the cognitive processes which are the common
focus of attention in current AI research—are computationally realizable, and *the wide claim*, that all mental states are so realizable.

It is argued that these are extremely different positions. The first
may ultimately hinge on little more than a decision over terminological usage. The second, however, is of ethical significance: the
mental states which are associated with consciousness or sentience,
particularly with enjoyment, suffering, etc., are ethically crucial; their
application does not rest on arbitrary decision. The narrow claim has
considerable plausibility in view of developments in AI. The wide
claim has little plausibility, it is suggested. Some bad arguments in
support of the wide claim are considered.

The chapter ends with some remarks on those aspects of
mentality which lie between the computationally tractable and the
computationally intractable—such as emotions, evaluative and ethical
choice. These and other areas of mentality, although 'non-cognitive',
may be capable of computational modelling. It is particularly

important that AI knowledge-based systems should be equipped with some kind of humanitarian normative outlook.

A FLOWERING OF ARTIFICIALITY

It has been a great summer for the artificial. Not merely were some five thousand delegates registered for the 1985 International Joint Conference on Artificial Intelligence in Los Angeles, but also Mrs June Tregale won a gold award at the Honiton County Show, Devon, for her recreation of an entire English country garden in plastic and silk flowers (*The Guardian*, 3 August 1985). A world first, surely. It would appear that the exhibit in question passed a floral equivalent of the Turing test: 'My biggest problem,' Mrs Tregale said, 'was the public. Half of them could not believe the flowers were fakes.'

If one is merely interested in exhibitions, then artificial intelligence and artificial flowers may serve as well as their respective natural correlates. Stick-in-the-soil horticulturalists may insist that something essential will always be missing from even the most delicately composed floral presentation. It is not clear that very much hangs on the issue.

In the case of intelligence, however, more may be at stake. On the one hand, AI supporters strenuously argue that as long as an artificially intelligent system *delivers* all (at least all) that a naturally intelligent system does, then that is all that matters as far as *intelligence* goes, because that is all that there *is* to intelligence. The view is not irresistable—indeed it has strenuous critics. Nevertheless, it has considerable plausibility. Enthusiasts of surrogate flora claim only that the latter may be *as good as* naturally grown varieties, and not that they *are* genuine flowers. AI supporters argue, however, that artificial intelligence *is* genuine intelligence: that human intelligence is merely the capacity of a naturally evolved organism to perform various sorts of operations—operations which are no different from those which might be performed by an artificial system.

On the other hand, intelligence is usually taken to be a feature of *mentality*: to be intelligent is to have a mind. To have a mind is to possess a lot more properties than merely those associated with the capacity for intelligent performances. Just what is involved in having a mind may differ from species to species (assuming that it is accepted that other species besides *Homo sapiens* have minds). In the case of humans, having a mind seems to involve, among other things, being able to have sensuous awareness of objects and of one's own experience and physical states; being able to experience pain and pleasure, love, hate, fear, anger, ecstasy, serenity; being able to be creative, inspired, nauseated, ashamed, bored; being able to tell jokes and find them funny, to play, to be aroused, to be satisfied, to suffer.

'INTELLIGENCE'

It is a highly *cognitivized* view of mind that emanates from AI and cognitive science literature. Most or all of the things mentioned in the above list no doubt involve the exercise of intelligence or cognition, and intimately so, and in a variety of different ways. Few of them are characterizable *exclusively* in terms of the exercise of intelligence—certainly not of 'intelligence' in the limited sense of the word with which we began: the sort of'intelligence which is concerned with performing tasks, solving problems, understanding, learning, rule following, discrimination, etc.

Often the term 'intelligence' is used in a rather looser and wider way than this, as more or less a synonym for 'mentality' as such. People working in AI often tend to be hazy about the relative boundaries of 'intelligence', 'mentality', 'thinking', etc. There is impressive historical precedent for such vagueness. In Descartes' philosophy, for example, any mental event is considered to be a modification of the *res cogitans*, the thinking substance which is one's soul. It has to be said here that verbs like *cogitare*, or *penser*, as used by Descartes, had an extremely wide range of application. '*Thought* (*cogitatio*),' he wrote, 'is a word that covers everything that exists in us in a way that we are immediately conscious of it. Thus all the operations of will, intellect, imagination, and of the senses are thoughts'.[1]

There is some suggestion here that for Descartes any mental event was a species of knowledge (of self-knowledge). He certainly believed that it intimately involved knowledge. Cartesian exegesis aside, the important point is to indicate how tempting it is to treat any process of mind as an act of thinking, a process of knowing or cognition. This cognitivizing tendency seems to be present in much subsequent philosophical or psychological writing. The matter becomes specially difficult and confused when 'cognition' is in turn understood in purely intellectual terms, so that any mental process whatsoever comes to be seen as an exercise of intellect, and therefore as something which is in principle susceptible of computational explanation within the AI paradigm. It is thus certainly very easy to slip from a narrow, relatively focused, sense of 'intelligence' to a wider, vaguer notion, a notion which seems to encompass the whole of mentality, including desire, emotion, direct conscious experience, pleasure, pain, etc. Because of this ready transition, the field of artificial intelligence seems to take on a much more portentious air. AI theorists become taken in by their own sleight of hand. We start out with a very limited and plausible enough claim concerning a particular set of cognitive activities—namely that computer performances of such activities may be called intelligent in exactly the same sense in which the human performances of those activities are. We end up with a grand theory of mentality as such—that any mental state, process, activity, capacity whatsoever is in principle computer simulable, and therefore computationally explicable.

THE NARROW AND WIDE CLAIMS

We must therefore distinguish two very different claims made within and on behalf of AI. There is *the narrow claim*, according to which that portion of human mentality which involves the exercise of intelligence (in some fairly well-bounded sense of the term) can be reproduced in working computer programs with complete fidelity, so that when a computer is displaying a certain kind of behaviour it is—*to this extent*—exemplifying genuine mentality. Then there is *the wide claim*, according to which any and all aspects of mentality can in principle be realized on computer systems of some arbitrary degree of complexity—or at least they can be explained in computational terms (in some pertinent, and non-trivial, sense of 'computational').

What I shall try to show in the following is the enormous difference in significance between these two claims; to show that they are not merely two variants, respectively cautious and incautious, of a common outlook. Much confusion has resulted from failing to distinguish these two claims and from failing to see how different they are. People who, for understandable reasons, find the narrow claim plausible, tend to think that there is relatively little extra cost involved in endorsing the wide claim. On the other hand people who, again for understandable reasons, are aghast at the wider claim therefore tend to turn their faces against considering the merits of the narrower claim.

There has been a lively debate of late concerning the merits of the narrow claim: Searle's arguments are intended to refute that version of the narrow claim which maintains that intentional mental predicates such as 'understands', 'means', etc., can be attributed to a computational system.[2] Now one of the crucial assumptions of Searle's position is that there is some 'essence' to 'meaning', 'intentionality' and other similar mental predicates and that this 'essence' governs in advance how such terms are to be attributed to appropriately performing computer programs, or the systems running them. Searle appeals to a philosophical tradition, inspired by Brentano,[3] which identifies intentionality as a defining mark of the mental. He wants to set up a firm boundary between primary notions of intentionality—'original' or 'intrinsic' intentionality, as he calls it—which cover the cases of meaning, understanding, planning, inferring, representing things by human beings; and all secondary or 'derivative' or 'observer-relative' intentional notions, covering the ascription of meanings, etc., to written or spoken texts, pictorial representations of various sorts, and, of course, crucially, the variety of intentional or quasi-intentional ascriptions which have come to be used more and more commonly in connection with various computational processes.

The need to set up this rigid barrier between 'original' and 'derivative' intentionality must be based upon the idea that something important depends upon this separation, but I wonder whether there really can be such an important dividing line here. A way to show that there might not be such a division after all is to contrast the question of boundary conditions for intentional notions with questions concerning notions of mentality where something important undeniably *is* at issue. Questions to do with sentience or consciousness—with actually directly experiencing things rather than simply meaning things—seem to be questions on which something quite practical does indeed depend. To decide that a certain system or organism possesses certain states of consciousness or sentient awareness may imply the adoption of a quite distinctive sort of attitude towards that system. To attribute intentionality to a system where such an attribution is made *in isolation*, i.e. in a manner such as to not imply consciousness or sentience as well, is not necessarily to become committed to that same sort of distinctive attitude. True, an enormous weight of theoretical, explanatory significance rests upon getting clear the nature of cognitive or intentional mental states. Supporters of the narrow computationalist claim may seriously underestimate the complexities of human cognitive processes, and therefore the extent to which they are open to computational analysis or replication. However, the issue of deciding on the merits of the narrow claim, considered in abstraction from the wider claim of *across the board* computer mentality, where this includes computer consciousness, seems really to be a matter of deciding how to extend an old classificatory scheme to new sorts of cases, rather than deciding whether some intrinsic property applies to the new cases.

THE NARROW CLAIM: SOME OPTIONS

Consider the following analogy. A sculpture may possess many features which its original possesses: for instance it may be exactly the same height as the human who is its model; it may have the same muscular outline, the same delicate shape to the hands, the same characteristic facial expression of exquisite sadness as is often found on its subject, and so on. Someone might object that the sculpture cannot really have a sad facial expression, since lumps of plaster cannot be sad, and only that which can really *feel* sad can have a sad expression. This would certainly be a possible way to determine linguistic usage, but it would be little more than that—a matter for verbal legislation.

Perhaps the person who is hesitant about ascribing sad facial expressions to sculptures is worried that to do so would be to end up having to talk of sad sculptures. The issue of whether sculptures can (not merely have sad expressions but) actually be sad is clearly not just a matter of verbal decision. Most people will agree without too much hesitation that a lump of plaster cannot possibly meet the preconditions of *being sad*, miracles aside. Only in a fairy story could the sculptor's art be fine enough to render the creation into a conscious individual.

The situation seems to me to be very similar with respect to computer simulations of the cognitive aspects of mentality. AI sceptics may say that no mere bunch of electronic circuits could genuinely possess intentional states such as understanding, intelligence, and so on—on the grounds that (1) only beings which are conscious could possess such states, and (2) a bunch of electronic circuits cannot be conscious. Now, let's concede (2), to cut short the argument, but why should (1)—the claim that only conscious beings can have intentional states—go unchallenged? Why, just because we ascribe states like understanding, etc., to appropriately functioning non-conscious systems, should we thereby be forced to ascribe properties such as consciousness to such systems—any more than we should have to say that the sculpture's sad face betokens a sad soul?

We now know that computers are capable of exhibiting performances—playing games, solving puzzles, planning, perceiving, etc.—which were hitherto thought to be exhibited only by creatures capable of undergoing genuine mental processes, genuine intentional states, such as understanding, etc. Some people claim that the success curve currently enjoyed by AI research is likely to flatten out soon and that it will turn out that only rather superficial and limited aspects of human cognitive activity are capable of computational simulation. Even if such AI sceptics turn out to be right, our conception of the nature of cognitive performances has been irrevocably changed by the displays of computer 'intelligence' that have already been achieved. We now know that, in these crucial respects, machines can *act* like us, even if it were to be the case that, as the AI sceptics insist, they cannot *mean* like us.

Given such computer performances, as currently achieved and as promised in the future, we have a choice. Here are some of the many possible things we could decide to say.

A first possibility is to say that computers, when exhibiting such performances, are not really playing chess, solving puzzles, etc., since doing those things involves mentality, which computers do not possess. What *they* are doing is *quasi*-chess playing, *quasi*-puzzle solving, etc.

A second possibility would be to agree that computers do play chess, solve puzzles, etc., after all, but to argue as follows: since these are the sorts of things which can only be performed by beings which undergo genuine mental processes and since, further, a computer would certainly not be thought to be living or conscious merely by virtue of being able to play chess or solve puzzles (for that would be a ludicrous jump), therefore some mental processes can inhere in entities which are neither alive nor conscious.

A third possibility would be to say that computers can have genuine intentional states, can be genuinely intelligent, but that such states are not by that token *mental* states, or are so only in a secondary sense. On this position intentionality, intelligence, cognition, etc., are not indissolubly linked with mentality. The study of cognition would then not be a study of mentality, but of some other category distinct, perhaps, from both the mental and the physical.

A fourth possibility would suggest that the question of what counts as a mental state is too undefined for beings which can only perform an extremely tiny subset of the sorts of things which people or higher animals can do with their minds, and that, whereas there would be little point in ascribing mental states to a machine which *just* played chess or *just* solved a limited range of puzzles, nevertheless it may well be possible to ascribe genuine intentional states to more advanced systems with a comprehensive cognitive architecture encompassing a great many human faculties, such as memory, language, perception, and so on. If such systems were sufficiently carefully designed, then it might be otiose to deny that they really understood, meant things, and so on. It may be that we would wish to make such intentional ascriptions in advance of admitting that such systems possessed consciousness or sentience.

Of the four, my own preferred option is the last one. However, as stated earlier, as long as we are limiting ourselves to discussing intentionality or intelligence or cognition, as opposed to consciousness or sentience, it matters relatively little which of these alternatives one chooses to adopt. What is missing in any discussion of intentionality or cognition in relation to computational systems is any *ethical* dimension, and it is the ethical dimension which is brought in when we discuss the wider claim about computer mentality. It is to the wide claim that we now turn.

THE ETHICAL SIGNIFICANCE OF THE WIDE CLAIM

It is the capacity to have these various features which we group together under the heading 'mental life', in the fullest use of that term, which we tend to regard as important in determining which entities or beings in the universe are potentially subjects of moral concern. Precisely because having a mind is, among other things, having the capacity to enjoy and to suffer, the notion of mind is central to our ethical thinking, for ethics is pivotally concerned with the provenance and distribution of enjoyment and suffering (if, no doubt, with other things as well).[4]

In order to shed light on this point, it would be useful to consider the ways in which people in our society are increasingly becoming concerned about the need to protect the interests of various species of animals. More and more people are coming to believe that domestic pets, farm animals, seals, dolphins, and so on, have quite elaborate mental capacities. While it is partly our view of the *intelligence* of such animals which is gradually being upgraded, it is also—and surely more crucially—our conception of the experiential or sensitive side of their mental life that motivates our increasing empathy towards non-human species. Computers or robots which exhibited mental capacities *only* of the cognitive kind—which were *only* intelligent (in the narrow sense)—would not, so far, be likely to excite our ethical sympathies a great deal.

Thus there are certain mental attributions which are theoretically and ethically low in cost: relatively little would be at issue if we conceded that a computer which exhibited cognitive performances such as chess playing, problem solving, sentence parsing, etc., was *to that extent* exhibiting genuine mentality. There are other mental attributions which are, from an ethical point of view, critical. If a computer or robot gave a convincing behavioural display of extreme suffering, it would be important to consider whether there was genuine suffering going on there as well—whether the outer display was indeed an indication of some genuine 'inner' state which rightfully ought to be an object of our moral concern. (It is not clear exactly what such a display would consist of, or *how* exactly it might succeed in convincing, but that is another matter.) In the case of states of consciousness or of experiential awareness—various sorts of pleasures and pains, for example—it is much easier to think of the mental state as conceptually distinct from the capacity to give the appropriate perform-ances than it is in the case of cognitive processes such as problem solving, task performing and all those other instances which lend themselves so readily to a computational treatment.

AI theorists are thus in a dilemma. If, on the one hand, they are claiming that all that is reproducible by computational means is intelligence (in the narrow sense) *without* consciousness, *without* any subjective experi-ence going on as well, then the debate over the genuineness or otherwise of artificial intelligence seems to be not much less sterile than that over the genuineness or otherwise of artificial flowers. If, on the other hand, they are agreeing that *genuine* intelligence involves (at least the capacity for) consciousness or subjective experience as well, *and moreover* that it is quite possible for a computational system to possess genuine states of consciousness, then they are making a much more adventurous claim—one which has much less intuitive plausibility and which is much more difficult to substantiate.

Many AI supporters tend to blur the distinction between intelligence and consciousness or to inflate the former notion so that it includes the latter as a necessary component. They do so at their peril. It seems (to me at least) difficult to deny the distinction between the capacity to exhibit intelligent performances on the one hand and the capacity to undergo conscious states on the other. To show that computers may generate the former is not to show that they may be subject to the latter.

CONSCIOUSNESS AND SCEPTICISM ABOUT OTHER MINDS

There are several highly questionable arguments which are often used to obscure this point. One such argument is the *appeal to scepticism about other minds*. The argument goes as follows. Just what is consciousness supposed to be, anyway? We do not (as individuals) even know whether our fellow humans have it, since all we can see is their external behaviour. So we would have as much or as little grounds for asserting it of an intelligently performing robot as we would for asserting or denying it of an intelligently performing human being, because the behaviour might be indistinguishable in the two cases.[5]

As against this, it has been fashionable, since Wittgenstein, to argue that the traditional philosophical doubts about whether fellow humans have a private inner consciousness accompanying their public behaviour is incoherent. Let us assume for the purposes of the argument that 'other minds' scepticism *is* coherent. The doubt about whether other human beings are conscious is a doubt about being able to validate a certain general belief, namely a belief that other individuals possessing *physiologies* just like mine, and exhibiting behaviour just like mine, also possess inner experiences like mine. The doubt about whether a computer can be conscious is of a quite different nature. It is a doubt over whether entities which have an entirely different physical structure, i.e. one comprised of electronic components and input–output devices of various kinds, rather than the wetware of a human or a mammalian central nervous system, can *also* possess consciousness.

Doubt about consciousness in computers (i.e. electronic devices of the sort that are currently used and that are visible on the research horizon) is an empirical doubt about whether such physical systems are capable of producing consciousness—whether they have the right 'causal powers', to use John Searle's phrase. It is a doubt which makes sense only in the context of the background assumption that certain physical systems *do* have the right causal powers to produce consciousness.

Moreover, given current knowledge in the neurosciences, it is difficult to see how such powers could be characterized except in terms of deep physiological properties of the central nervous systems of humans and other higher animals. That is not to say that, in holding such an assumption, we are committed to believing that *only* beings with central nervous systems like ours can have consciousness. It may well be that the causal powers which enable our central nervous systems to produce consciousness can also be possessed by physical structures of a quite different kind. There may be all kinds of sentient extraterrestrial beings with (to us) highly exotic neuroanatomies.

The classical philosophical doubt about other minds, on the other hand, is a doubt of a quite different kind. It expresses a despair over being able to be sure of (or even, more radically, over being able *to make sense of*) any causal generalization of all concerning the physical bases for consciousness. It is, in short, a doubt that has to be put to rest before the neurosciences can begin and before any speculation over consciousness in non-human kinds of physical systems can begin (and, of course, before ethics can begin).

COMPUTATIONAL AI VERSUS 'PSYCHOTECHNICS'

The issue of attributing consciousness or mentality to computational systems is further clouded by an unfortunate ambiguity in the notion of 'artificial', as used in the context of the phrase 'artificial intelligence'. The latter term was coined in the mid-1950s to characterize a sub-branch of the infant field of computer science.[6] The term was and is intimately associated with intelligence or mentality in computing machines, but there may well be many other ways of producing artificial intelligence or mentality apart from designing or programming computers.

It is at least a remote empirical possibility, for example, that one day a means could be developed for replicating the biological structure of an entire living, breathing, thinking and feeling human being—or at least of some lesser organism which nevertheless had some vestigial sentient states. If, to take a science-fictional point of view for a moment, one discovered that the person with whom one had just enjoyed a long conversation (or love affair) was an artefact, one would not necessarily withdraw one's attribution of consciousness. If it turned out that one's companion really did have a central nervous system just like one's own, for instance, then the fact that it was an artificially produced one need not make any difference. If processes of consciousness such as pain and pleasure are indeed produced by *states of the central nervous system*, then an artificial central nervous system may still produce real pain or pleasure. That is, it would have the appropriate 'causal powers', but these causal powers would in this case be the result of fabrication rather than natural growth.

In discussing the possibility of such artificially produced conscious beings, with real feelings, desires, interests, etc., I am not implying that it would be in any way *morally desirable* to bring beings of such a kind into existence. Indeed, the idea might strike some people as quite obnoxious. This raises many interesting issues, but it is beyond the scope of the present discussion to enter into them here. I am presently concerned only with the fact that a special ethical dimension does attach to their creation which is missing from the narrow AI activity of creating cognitively acting systems. The latter, unlike the former, will not have *interests*—not, at least, in the full sense in which human beings have interests. Here, once again, we have a pressure to make a distinction between primary and derivative attributions. It is a crucial part of my argument that, in the case of attributions of *interests*, it is of great importance where we site that division; this is in marked contrast to the question of where to site the boundary between primary and derivative attributions of intentionality, cognition, etc.

There is also, of course, the question: 'What counts as an artefact, as an artificial, as opposed to a natural, X?' There are no doubt many actual or potential ways of producing biological organisms which have varying degrees of artificiality. Are cells synthesized in a laboratory artefacts? There is clearly a range, or a series of ranges, of intermediate cases between the paradigmatically artificial and the paradigmatically natural. A purist would doubtless object that not just Mrs Tregale's but all the exhibits at the Honiton County Show were artificial, being the products of human cultivation.

So it is not the *artefactuality* of computers which makes it difficult to envisage consciousness in them. There might be all sorts of artificially created beings which possessed 'inner' states of a kind that would be psychologically and ethically interesting. However, consciousness being the sort of thing it is, such artefacts would no doubt be extraordinarily complex things—much more complex than fifth, sixth or seventh generation *computers* are likely to be, surely. Of course the field of artificial intelligence, taken in its widest sense as the quest to build artificial systems possessing genuine mental states of all kinds, need not be dependent upon digital computer technology alone. Moreover, who knows how computer technology is going to evolve and with what other technologies it will merge—either those now developing in parallel or other technologies not yet in existence.

Thus AI in this wider sense—*psychotechnics*, as it might be called—may, in decades or centuries to come, succeed in building artificial minds in the fullest and most ethically meaningful sense. No doubt computational systems, or their distant descendants, will also play a central role in any such future psychotechnologies, just as, no doubt, computational principles play a central role in the functioning of our own psychophysical organization.

THE SUPPOSED ETHICAL STATUS OF ATTRIBUTIONS OF CONSCIOUSNESS TO MACHINES

Another argument which is used to broaden the claims of AI (in its narrow, computational sense) might be called the *ethicization of attributions of consciousness*. Sloman, for example, claims that the general question of whether computers (of a particular complexity of organization) might ever be genuinely conscious is fundamentally ethical in nature, since it is dependent upon how we choose to treat such machines.[7]

If Sloman were right, then it would become easier to see the possibility of a *global* account of mind based on purely computational principles. With growing sophistication of AI techniques and associated hardware design, computers and robots will no doubt display progressively richer and more varied capabilities. On Sloman's view, the question 'Are they really conscious?' would be answered by our *practice*, by the fact that we would be unable to treat them in any other way than *as* conscious, i.e. as appropriate subjects of moral concern.

According to this view, then, it is not the case that we adopt an ethical attitude to a certain class of being *because* we believe them to be capable of certain sorts of conscious states—a belief which may or may not correspond to some independent factual state of affairs. Rather is it that our belief that they have those conscious states *consists in* our adopting the ethical attitudes. This would seem to imply, among other things, that there would be no possibility of our being systematically in error—of our treating *as* sentient systems that were not, or vice versa. As Sloman put it in the closing passage of a paper read to this year's International Joint Conference on AI:

> When we have shown in detail how like or unlike a human being some type of machine is, there remains a residual seductive question, namely whether such a machine really can be conscious, really can feel pain, really can think, etc. Pointing inside yourself at your own pain (or other mental state) you ask 'Does the machine really have *this* experience?'. This sort of question has much in common with the pre-Einsteinian question, uttered pointing at a location in space in front of you: 'Will my finger really be in *this* location in five minutes' time?' In both cases it is a mistake to think that there really is an 'entity' with a continuing identity, rather than just a complex network of relationships. The question about machines has an extra dimension: despite appearances, it is ultimately an *ethical* question, not just a factual one. It requires not an answer but a practical decision on how to treat the machines of the future, if they leave us any choice.[7]

In this passage, Sloman appears to be making two different points. The first is that there is a certain sort of question, 'Can such and such machines really be conscious, feel pain, etc.?', which appears to be purely factual but which is in fact ethical (or partly so). The second is that the question is, like the pre-Einsteinian's question about absolute points in space, a pseudo-question. These two points are meant to give mutual support to one another, but they seem to be mutually contradictory.

In order to see this, consider our earlier distinction between narrow AI, which makes use exclusively or centrally of computational organizations and techniques, and wide AI, or psychotechnics, in which computational technology is enhanced by other, perhaps as yet undreamt-of, technologies. As I said before, there seems to be no compelling reason in principle why artificial systems should not be created which possessed the capacity for consciousness, pain, etc. Take, now, some hypothetical future successful product of psychotechnology: some artefact which reproduces a wide range of behavioural and physiological features which we take to be determinative or constitutive of possessing a mental life. (We leave it an open question as to what role AI technology, as viewed through today's eyes, plays in our hypothetical artefact.) Suppose someone were to say, of such an artefact: 'Yes, it appears to behave in ways which indicate consciousness, sentience, and so on, and it incorporates all those casual properties which we believe on the best available evidence to be responsible for consciousness in ourselves. But is it *really* conscious? Is it really the same sort of thing as *this* that I feel and know to be my subjective experience?' Clearly, such a question could be considered to have a similar sort of incoherence as that which affects the pre-relativist's query about absolute points in space.

It could also be considered incoherent for a different, more straightforward reason, for we are assuming that we have an artefact which is believed, on the best available evidence grounds, to have all the necessary physical requisites for possessing genuine consciousness. To accept all this, but to raise the possibility that it might not be conscious, is surely simultaneously to accept and call into question a certain theory of causally sufficient conditions for consciousness. In other words, it has appropriate consciousness-endowing causal powers, but perhaps in this case the powers are that having their effect.

Clearly there is nothing in the incoherence of such a question which suggests that the issue is an *ethical* one, in the sense of 'merely a matter for ethical decision'. Indeed, if the question 'But is it really conscious?' really is an incoherent question in this setting, then surely it can not also be an ethical one. I take it that ethical issues are not, *per se*, incoherent issues. (It would certainly need very special arguments from Sloman to show that they were.)

I think Sloman has failed to analyse the structure of his own ethical thinking properly. I take it that Sloman endorses the same general humane moral principles as are shared by most educated people of our culture, in respect of avoiding suffering, seeking to alleviate it where possible, etc. Such a principle would, of course, not be limited to *human* suffering: most people recognize at least some non-human suffering as ethically of concern (e.g. they are shocked by cruelty to pets, etc.). There might, of course, be *practical* disagreements, e.g. between meat eaters and vegetarians on whether, and if so, how much, pigs or chickens actually do suffer. However, usually the differences here are factual. Certainly the people who might be in strong ethical disagreement, in the sense of having quite unreconcilable views on what ought to be done in a given case, may share all actual ethical beliefs in common and just disagree over certain factual matters. Suppose, now, that Sloman and I had such a practical dispute about a particular candidate 'psychotechnical' being—he claiming that it was suffering and therefore that it merited our assistance and I claiming that it was not and thus did not. Our *ethical* beliefs may well be quite in harmony: they certainly *need not* conflict with each other. It would be sufficient to cause such a practical disagreement for us to share some general ethical principle of the form:

(P) Whenever conditions C obtain then (other things being equal) action A should be performed.

Sloman and I would then be involved in a *practical* dispute because we disagree over whether A should be done, but our disagreement is over whether or not conditions C obtain. The question of whether conditions C obtain is not itself an *ethical* question. It is a factual question—it *has* to be, otherwise the general principle (P) (which, we are supposing, Sloman and I share) would be incapable of being properly framed.

THE 'TRUTH' ABOUT CONSCIOUSNESS ATTRIBUTIONS

In any case, surely it is abundantly clear that the general issue of consciousness in artefacts *is not* an ethical one (in Sloman's sense) but a factual one. Of course there are difficulties in directly checking that a given artificially produced system has direct states of consciousness, but we could have strong evidence of various sorts. The evidence would not just be behavioural. In order for any psychotechnological initiative to be successful, an enormous corpus of knowledge concerning the physical bases of conscious states will have to have been built up. It is obvious that today's neurophysiological knowledge is only at a relatively primitive stage compared to the kind of knowledge that will be needed in order to build a biological system capable of supporting consciousness. However, there seems no reason why that knowledge should be any less factual in nature than any other corpus of scientific knowledge.

There might be quite outlandish forms of empirical testing for the presence of conscious states in an organism, natural or artificial. It may turn out, for example, that certain people have 'ESP'-like capabilities that correlate well with human pain. That is, they may be able to identify faultlessly whenever a human being in the next room is in pain or not. They may be able to perform similarly with animals. Why should they not also be able to identify similar states in robots or other artificial beings? (Sceptics may fail to be convinced by the evidence of such specially gifted people, but the question at issue is the factual *status* of claim that such artificial beings might be conscious, not the ease with which people might be convinced of such a fact, if indeed it obtained.)

It is often suggested that, the progress of research in neurophysiology, medicine, etc., terms like 'consciousness', 'pain' and other familiar notions of our intuitive folk-psychology will be shown to be too primitive and contradictory to be of use in a properly scientific account of mental life. Dennett has argued for this view most forcefully in his paper 'Why you can't make a computer that feels pain'.[8] This would give an additional reason for doubting the factual nature of questions like 'Is such and such an artefact really in pain, really conscious?'.

Wilkes has pointed out, in a recent article, many difficulties for a simple-minded application of the notion of pain. A striking instance is provided by the case of hypnotic anaesthesia. She cites experimental work of E. R. Hilgard:

> A typical experiment is the following: subjects are hypnotized, told they will feel no pain, and then one arm is put into a stream of circulating iced water. This is rapidly experienced as unpleasantly painful by the unhypnotized; but subjects under hypnosis may sincerely report that they feel no pain, and will leave their arm in the water for long periods apparently untroubled. On the other hand, and *with* the other hand—if it is supplied with pencil and paper—the subject typically provides a simultaneous running complaint about the intensity and unpleasantness of the pain.[9]

Such cases simply show that there are occasions when the conditions of the application of the predicate 'is in pain' are not straightforward. They do not show that the predicate is incoherent or scientifically invalid, nor do they show that it is not a factual predicate like 'is in the bath'. Such arguments surely do not threaten central uses of such terms. They do not imply that we should take our general ethical concerns over the occurrence of pain (in the vast majority of cases which do not display such anomalous features) less seriously.

CONCLUSION: AI AND ETHICAL MODELLING

Thus the question of consciousness, pain, etc., in artefacts is a genuine one—a 'factual' one. It is not simply a matter for ethical decision. Nevertheless, as we have seen, it is not an issue to which ethical considerations are indifferent. If one day we are indeed in a position to produce artefacts which do appear by all available tests to be conscious, then we will be saddled with some important new ethical problems. We will have to worry about their interests as well as our own. Genuinely intelligent and sentient artefacts buried beneath the rubble of an earthquake, for instance, would, if still 'alive', no doubt have a direct claim to be rescued, just as would human or animal victims. If their sentience and thus their capacity to suffer is genuine, then our general obligations to alleviate suffering wherever possible would, in consistency, have to be extended to such artefacts. Merely intelligent, and non-sentient, robots, however, would not necessarily prompt the same direct obligations, although they might need to be pulled out from the rubble as important scientific instruments.

I have also suggested that genuine consciousness is not likely to result merely from more and more sophisticated *computational* developments. Debates about whether *computers* are ever likely to have consciousness, pains, emotions, etc., seem to me to be sterile, insofar as computational devices are understood in their present-day sense as machines which process sequences of symbols. It may be possible to produce computational systems which *model* various aspects of consciousness, emotion, and so on, but a computer running a program which models consciousness is clearly a quite different thing from a computer which, by virtue of the program which it is running, *is* conscious. There is certainly no *a priori* reason why the latter should occur.

Thus while there is no doubt that a good proportion of our mental capacities are based upon symbol or information processing, and are therefore explicable computationally, it seems equally clear that the ethically crucial aspects of our mental life rest upon quite different sorts of properties. We have, at present, only a very hazy idea of what those properties might be. Perhaps such states of qualitative awareness may involve computational factors in their explanation, but their explanation cannot be *exclusively* computational.

The computational paradigm is thus unlikely to offer a *global* account of 'the mind'. To say this is to belittle AI hardly at all, since it is clear that AI has a fundamental role to play in explaining mentality. Work in AI has to date been very largely concerned with mental processes which are characterizable as 'cognitive' in the narrow sense of having to do with task performance, rule following, factual thought and discourse, instrumental or means-end planning, perceptual processing, and so on. Some speculative work has been done on producing computational models for emotions, aesthetic and evaluative attitudes, choices, etc.,[10] but this has very much been a Cinderella area of AI. This is partly, no doubt, because of its relatively low payoff in military and commercial terms, but it must also be due to the sheer *difficulty* of getting a theoretical focus on, say, emotion as opposed to cognition.

The interesting thing about emotions, desires, etc., is that they seem to contain both cognitive and qualitive aspects inextricably linked. To feel anger, or jealousy, or sexual excitation is both to experience certain bodily sensations and to make certain judgements or classifications. The mental and the physical aspects of emotional states seem to be intertwined (the 'hot' ones, at least, as opposed to 'cool' ones, such as affection, concern, etc.). Within the mental side of, say, fear, one can discern both a purely qualitative side and a cognitive side. Being afraid seems to feel like something 'on the inside', rather like being in agony from toothache; but, unlike the latter, it will usually turn out to involve a complex structure of judgements which wire into a person's general belief and value system (I am afraid that X might happen because, if it does, then Y will occur, which will diminish the possibility of Z, which I both greatly desire and believe myself to need . . . , etc.).

The cognitive aspects of our emotions and attitudes—which are the aspects that lend themselves to computational modelling—therefore cannot be underestimated. That applies in particular to our moral emotions and attitudes, righteous indignation, remorse, admiration, as well as to our more rational moral judgments concerning what it is right to seek and to avoid, concerning the relative merits of competing claims or norms, the apportionment of responsibility, and so on.

It is not clear whether any such computational ethical modelling is likely ever to be either convincing or useful, but it may be that only through becoming actively concerned with building an explicit ethical orientation, and moreover a humanitarian ethical orientation, into AI systems that there is any hope that the AI technological paradigm can maintain any pretence at being a humanizing influence in our civilization. The main volume of research in the field of AI is becoming more and more subservient to the needs of developing advanced systems of weaponry and warfare, to the accelerated accumulation of wealth and power by multinational industrial concerns, to the battles for world market domination between the economic superpowers. AI seems to be getting increasingly alienated from that fascination with understanding the complexities of human thought processes which was once its guiding inspiration.

There are many things that can contribute towards keeping this spirit of fascination alive. One activity which might help to combat the excessively instrumentalistic approach to thinking which dominates AI at present is that of trying to build moral norms—and appropriate ones—into knowledge bases. It looks as though we are entering an age in which electronic 'intelligent' knowledge bases will increasingly be considered as oracles to consult and to defer to, as dominant repositories of Truth. If these oracles are to serve the interests of human beings around the world, rather than merely the interests of the Fortune 500, then they must be given more than merely domain expertise: they must be provided with some measure of social and normative enlightenment. It remains to be seen if the moral wisdom of ordinary humankind is too ineffable, too inscrutable, too variable, to be captured within an AI representation. We can only hope that the attempt to inject human responsibility into our artificially intelligent systems manages to succeed.

ACKNOWLEDGEMENTS

Discussions with very many people helped to form and clarify the ideas presented herein. I would like to thank the following: Margaret Boden, David Conway, Jon Cunningham, Steve Draper, Peter Forte, Andre Gallois, Andrew Redfern, Guy Scott, Blay Whitby, Masoud Yazdani, and, especially, Richard Hare and Aaron Sloman.

REFERENCES

1. Descartes, R. (1967). *The Philosophical Works of Descartes* (translated by E. S. Haldane and G. R. T. Ross), vol. II, Cambridge University Press, p. 52 (Appendix to Reply to Objections).
2. Searle, J. (1980). 'Minds, brains and programs', with Open Peer Commentaries. *Behavioural and Brain Sciences*, **3**, 417–457.
3. Brentano, F. (1960). 'The distinction between mental and physical phenomena'. Translation by D. B. Terrell of a section from F. Brentano, *Psychologie vom empirischen Standpunkt*, Vienna, 1874. In *Realism and the Background of Phenomenology* (Ed. R. M. Chisholm), Free Press, Glencoe, Illinois.
4. The question of whether there is any necessary *content* to ethical thinking has been much discussed within moral philosophy. In my doctoral thesis (Torrance, S., 1977, *Non-descriptivism: A Logico-Ethical Study*, DPhil Thesis, University of Oxford) I proposed a formalistic account of ethics while arguing that the formal structure of ethical thinking constrained rational ethical thinkers to be centrally concerned with human welfare and harm (see Hare, R. M., 1963, *Freedom and Reason*, and 1981, *Moral Thinking: Its Levels, Method and Point*, Clarendon Press, Oxford). The same arguments would apply to concern for the welfare and harm of non-humans, including artificial beings, if these latter genuinely could suffer and enjoy things, as well.
5. See, for example, Turing A. M. (1950). 'Computing machinery and intelligence', *MIND*, **LIX**, 433–460.
6. The first published appearance of the term 'artificial intelligence' is believed to be John McCarthy's proposal that 'a two-month, ten-man study of artificial intelligence be carried out during the summer of 1956 at Dartmouth College in Hanover, New Hampshire. The study is to proceed on the basis of the

conjecture that every aspect of learning or any other feature of intelligence can in principle be so precisely described that a machine can be made to simulate it.' Quoted in Charniak, E., and McDermott, D. (1985), *Introduction to Artificial Intelligence*, Addison-Wesley, Reading, Mass., p. 11.

7. This claim can be found in many of Sloman's writings: the most recent source is Sloman, A. (1985), understand?', *Proceedings of the Ninth International Joint Conference on Artificial Intelligence*.

8. Dennett, D. (1978). *Brainstorms: Philosophical Essays on Mind and Psychology*, Harvester Press, Brighton, Chap. 11.

9. Wilkes, K. V. (1984). 'Is consciousness necessary?'. *British Journal of the Philosophy of Science*, **35**(3), 223–243.

10. See, for instance, Sloman, A., and Croucher, M. (1981), 'You don't need to have a soft skin to have a warm heart: towards a computational analysis of emotions', Cognitive Studies Research Paper CSRP 004, University of Sussex, Brighton.

Artificial Intelligence for Society
Edited by K. S. Gill
© 1986 John Wiley & Sons Ltd

7. AI AND PHILOSOPHY: RECREATING NAIVE EPISTEMOLOGY

JANET VAUX Freelance journalist, Editor, *Machine Intelligence News*

'Reasoning is but reckoning,' said Hobbes in the earliest expression of the computational view of thought. Three centuries later, with the development of electronic 'computers', his idea finally began to catch on; and now, in three decades, it has become the single most important theoretical hypothesis in psychology (and several allied disciplines), and also the basis of an exciting new discipline called 'artificial intelligence'[1]

If Hobbes was, as Haugeland suggests, three hundred years ahead of his time, what does that imply about the direction that philosophical epistemology has taken in the three centuries between Hobbes and the 'rediscovery' of the computational view of thought? Haugeland's answer is that it was sidetracked by a number of philosophical dilemmas—including the problem of dualism—which, he seems to suggest, a computational approach can in fact surpass: 'Cognitive scientists can be materialists (nondualists) and mentalists (nonbehaviourists) at the same time; and they can offer explanations in terms of meaning and rule-following, without presupposing any unexplained homunculus. It all depends on a marvellously rich analogy with computers. . .'.[2]

Some philosophers may be sceptical that the hoary old problems have been so easily brought to a solution—particularly one that is at once materialist, mentalist and non-dualist. Others may suspect that the appeal to Hobbes is a symptom of the fact that AI theorists—and sympathetic philosophers—are merely reinventing the wheel, epistemologically speaking. Are such philosophers simply trying to defend their historical territory, and how much can AI really learn from philosophy?

The ultimate goal sometimes claimed by AI theorists—that of recreating human intelligence—certainly appears to raise a number of philosophical issues. Indeed, it raises these issues all too easily: the question of whether a machine could think is all too familiar to a philosopher, engendering a range of problems including the mind/body problem, the other minds problem and the various problems of consciousness, intentionality, subjectivity, and so on. There is a question of whether the AI theorists are really interested in these issues in quite the same way as philosophers are. Whether a machine can think delimits a practical problem (albeit a long-term one) for an AI theorist; it tends to invoke an exercise in conceptual and analysis and logical possibility for philosophers in the Anglo-Saxon traditions.

To illustrate this, we will look at John Searle's much disputed Chinese Box puzzle—not in order to attempt a new solution but to consider it as an example of failure to communicate between AI and philosophy. Searle says:

> Suppose I am in a closed room and that people are passing in to me a series of cards written in Chinese, a language of which I have no knowledge; but I do possess rules for correlating one set of squiggles with another set of squiggles so that when I pass the appropriate card back out of the room it will look to a Chinese observer as if I am a genuine user of the Chinese language. But I am not; I simply do not understand Chinese; those squiggles remain just squiggles to me.[3]

And so it is with a digital computer; it simply has the rules for correlating squiggles.

I think Searle makes the point that he sets out to make here: he shows that there is a meaningful distinction between understanding (or language use) and the simulation of understanding or of language use. He does it by an established method of analytic philosophy—the thought experiment (or science fiction tale), attempting to stretch the application of a concept beyond its meaningful limits. The AI community is not in the least impressed. One gets the impression (borne out at least by Searle's account of the various responses to him[3]) that his AI protagonists have really failed to engage with this particular argument of Searle's.

A reason for this, I suggest, is that the example is a little bizarre from an AI point of view in that it concedes from the start that a perfect simulation of language use is possible. This must surely be what is at issue. If a machine could be constructed that behaved in every way as if it could understand language—if there was literally nothing to which you could point and say, that's how you can tell it's not really understanding—then Occam's razor would suggest that Searle's understanding of 'understanding' is superfluous, however well established in the culture. If it is the current concept of 'understanding' that you are trying to explicate, then Searle's arguments are telling.

Of course, I have taken Searle's example out of the context both of his own broader arguments against thinking machines and of the particular target of the Chinese Box story—those AI enthusiasts who claimed that primitive 'natural Language' systems such as Winograd's SHRDLU already displayed understanding of a sort comparable to human understanding.[4] In fact, this example is quite frequently argued simply on its own merits, and my point is that the development of this particular sub-debate highlights a significant disparity of concerns between AI theorists and philosophers.

A similar pattern of debate can be seen on other philosophical topics, such as intentionality or consciousness, where from the AI point of view there is no immediate mileage in any concepts that can not be explicated in functional definitions. It seems, for example, to be common wisdom among AI people to remark that it is an open question whether or not intelligent machines will be conscious—virtually an admission that consciousness does not make a lot of sense from an AI point of view (but if it is functionally related to intelligence then it should somehow show up in the simulation eventually). Others in AI simply decline to talk about a concept for which there is no agreed definition.

One of the problems of consciousness, regarded from the standpoint of strict scientific standards, is that the only evidence for it is subjective. This is one basis of the so-called 'other minds' problem, the question of how I can know that other people also experience the world as I do—that they are also conscious subjects. This philosophical chestnut is seized on by many in AI as a way of avoiding the question of whether an intelligent machine would be conscious: if we do not even know whether (other) people are conscious then we shall not know with machines either and it does not really matter. This is a bit disingenuous, for the philosophical problem does not actually imply a genuine doubt about other people; it is formulating a problem about accounts of knowledge based in standards of scientificity that are arguably the wrong sorts of standards to apply to our direct knowledge of other people. We do not actually hypothesize the existence of other subjects on the basis of (an illicit) generalization from our own case! We *know* other people exist, however hard it may be to provide an epistemology.

One price that is paid for the failure to communicate is mutual irritation. There is a sense among AI people that philosophers who raise objections to their project are caught up in the cobwebs of dualism and various metaphysical concepts such as intentionality (or some other mysterious ingredient X). In contrast AI appears to be striving towards scientificity, its materialism demonstrated in the hypothesis of human beings (or perhaps human brains) as machines.

The trouble is that dualism cannot be rejected simply by an act of will; it is not enough to say 'I don't believe in ghosts, or in souls' (souls are not, in fact, a central philosophical issue). In practice, if the AI perspective does surpass dualism, then it is in the direction of a thorough-going mentalism, where the formal system reigns. The 'materialism' of AI seems to consist either in the basis that the ultimate dependence of thought on the brain must be explicable (which seems unobjectionable) or perhaps in a presumed analogy between the functioning of brain synapses and of binary connections in a computer. Neither of these theses actually does any work in the computational approach to cognition—or at least in mainstream versions.[5] Indeed, it is generally claimed that the machine (brain or computer) is irrelevant to the validity of the system.

I am not sure, anyway, whether the question of dualism is really the philosophical ground on which AI should attempt to take a stand. It is so hard to resist the hardware/software model in any sort of computational analogy, and this provides a rather unoriginal form of dualism, with all its attendant problems (e.g. the 'link' between hardware and software in computers is achieved by electronics engineers and programmers; how is the analogy supposed to operate on this point?).

More problematic is AI's commitment to formalism. This is, arguably, essential to AI, which has its roots in mathematical logic, and there is no reason to suppose that the full potential of formal reasoning systems for automation and mechanical control has yet been sighted. However, as a model of intelligent human behaviour it can run up against difficulties.

In this respect it is interesting to look at some of the areas were AI is running into practical (as opposed to metaphysical) problems. While computers can beat the average human being at 'clever' tasks such as playing chess and proving theorems, they are impossibly stupid over the sorts of tasks that most of us take for granted—like stacking bricks, speaking our native language, or finding the way to the door. If human beings really moved across a room by means of suppressed algorithms (or even rough and ready rules), then it should be easier than it is to simulate this sort of intelligent behaviour.

The work of the French philosopher Maurice Merleau-Ponty,[6] which locates our knowledge of space directly in our bodies and our business in the world, is surely more phenomenologically accurate than the lonely brain of AI theory, processing signals from a distant world by means of the inadequate rules of 'naive physics'. Again, Wittgenstein's arguments that the meaning of language must be based in social use and a community of users are worth rereading in the light of the computational analogy: how can the computer have this sort of direct access to language, unmediated by machine code?[7]

The assumption of many AI theorists seems to be that it is just a matter of reproducing the rather unscientific mess of ideas that guide us round the world and through our relations with other people—of recreating 'naive physics' and 'folk psychology'. The 'obviousness' of this approach to epistemology is no guarantee to its accuracy. There is a danger simply of recreating naive epistemologies.

Finally, while the idea of simulating human intelligence may have been the historical inspiration of AI, it does not define the discipline as it exists now (apart, at least, from the related field of cognitive psychology). Any philosophical contributions about the nature of human intelligence only matter where they imply some sort of practical difficulty in the creation of machine intelligence (a different matter). It is interesting, incidentally, that the rather ambitious claims made for some of the apparent natural language 'breakthroughs', such as Winograd's SHRDLU[4]—one of the targets of philosophers such as John Searle and Hubert Dreyfus—are now recognized by natural language researchers to have been overoptimistic.[8] Dreyfus' persistent—phenomenological based—critiques of current natural language work seem also to have had a considerable impact on Winograd and others.[9]

Nonetheless, AI is proving itself pragmatical enough to attract major funding from governments, the military and industry. However, if it is coming up with the goods, those goods are not androids.

REFERENCES

1. Haugeland, J. (Ed.) (1981). MIT Press, p. 1.
2. Haugeland, J. (Ed.) (1981). *Mind Design*, MIT Press, p. 5.
3. Searle, J. (1980). 'Minds, brains and programs'. *Behavioural and Brain Sciences*, **3**, 417–457; also reprinted in Haugeland[1] and also *Minds, Brains and Sciences*, BBC, 1984.
4. Winograd, T. (1972). 'Understanding natural language'. *Cognitive Psychology*, **1**.
5. This is perhaps not entirely true of some more adventurous projects, such as the Bolzmann machine, associated with Professor Geoffrey Hinton of Carnegie Mellon University.
6. See in particular, Merleau-Ponty, M. (1962), *The Phenomenology of Perception*, (English translation by C. Smith), Routledge & Kegan Paul.
7. *Wittgenstein, L., Philosophical Investigations*, Basil Blackwell, 1953.
8. See Ritchie, G., and Thompson, H. (1984), 'Natural language processing', in *Artificial Intelligence* (Eds T. O'Shea and M. Eisenstandt), Harper and Row.
9. This will apparently be evident in Flores, F.C., and Winograd, T. Understanding computers and cogrution, in Winograd, T. (1983) Language as a Cognitive Process, Volume 1: Syntax. Addison-Wesley, New York.

Artificial Intelligence for Society
Edited by K. S. Gill
© 1986 John Wiley & Sons Ltd

8.

DEVELOPMENTAL PSYCHOLOGY'S CONTRIBUTION TO COGNITIVE SCIENCE

JULIE C. RUTKOWSKA Cognitive Studies Programme, University of Sussex, Brighton

ABSTRACT

Developmental psychology can make a unique contribution to cognitive science, and its low status in relation to AI is unjustified. This relationship is discussed in the context of growing criticism of the computational metaphor for the mind and the limited results from machine learning.

AI 'functionalism' implies that the mind can be studied by constructing computer programs. The organization of infant abilities emphasizes that the function of the mind is action situated in the world. We need to redraw the boundaries of cognition and computation to include behavioural processes and the environment. The abilities of naturally intelligent systems rely on an externalization of processing which exploits and is scaffolded by physical and social environmental structure.

AI concentrates on 'task-oriented learning'. Microanalytic developmental studies reveal similar patterns of restructuring in different age periods and domains of activity. These suggest the need for a theory of 'knowledge-oriented learning'.

INTRODUCTION

This paper is concerned with developmental psychology and with that side of artificial intelligence (hereafter AI) which aspires to model human abilities. Informed use of computers for educational and other purposes must be supported by a principled theory of cognition and of its development, and many trust that the emerging interdisciplinary field of cognitive science will provide us with such a theory. Although cognitive science purports to draw upon AI, linguistics, philosophy and psychology, it has shown scant interest in developmental psychology. The present paper is motivated by the grossly asymmetrical status of AI and developmental psychology within cognitive science. It will be argued that both the study of particular developmental periods and the comparison of developmental processes across different ages and domains of ability can make a unique and necessary contribution to the cognitive science endeavour. Some of the difficulties currently confronted by AI might stand a greater chance of creative resolution if it took note of research in developmental psychology.

AI AND DEVELOPMENTAL PSYCHOLOGY IN COGNITIVE SCIENCE

Cognitive science is not yet a unified area with a clearly defined paradigm constraining its theoretical constructs and research methods, but some general areas of consensus can be drawn out. Contributory research shares two main characteristics: it aims to provide us with a deeper and more precise understanding of the *mind* than we have so far achieved and it believes that the kinds of intelligent abilities which are evidence of the mind are made possible by the subject's *knowledge* of the world. There is a commitment to conceptualizing the mind and knowledge in terms of some version of the *computational metaphor*. The mind is viewed as a physical symbol system—a physical machine which can construct symbols referring to objects and events in its external and internal environments. Knowledge is viewed as an active process of computation or rule-governed symbol structure manipulation.

It is easy to see how AI, with its links with computers and computation, has gained such a central place in cognitive science. Essentially, AI attempts to model our abilities by programming computers to perform them. The general argument for its relevance goes something as follows. A computer is a type of physical symbol system because locations or addresses within it can be considered as abstract symbols to which researchers can assign particular meanings. An analogy may be drawn between the relationship of a computer to its *program*—the structure controlling which symbol manipulation processes it performs and the order in which they are activated—and the relationship between the brain and cognitive or mental processes. A single program can be run on many computers with different physical or hardware characteristics, supporting the view that symbol structures and their manipulation can be understood independently of the particular physical machine in which they are implemented. Thus, it follows that the significant features of the mind can be modelled by a computer program despite the very different physical characteristics of brains and computers—one variety of the philosophical doctrine known as 'functionalism'.

AI and cognitive science in general share a methodological emphasis on what Miller calls 'theory development',[1] the construction of detailed models of mechanisms which are sufficient to produce the phenomena of interest. In principle, a working program can operationalize mentalistic constructs and establish that their interaction could actually lead to a given outcome. In contrast with this, mainstream developmental psychology proceeds within a low status 'theory demonstration' framework. It concentrates on the subject's successes and/or failures in task situations where the outcome is supposed to establish the relative validity of what may be equally inadequate theories (compare Johnson-Laird's comparison of cognitive science and psychology in terms of their respective adherence to the 'coherence' and 'correspondence' theories of truth[2]).

The developmentalist who queries the validity of AI theories exposes themself to the challenge that they are better than no theory at all. In many respects, this challenge is appropriate and well founded. However, does mainstream AI live up to the expectations one might form by contrasting it with mainstream developmental psychology? In several important respects the answer to this question has to be 'no'. The main part of this paper discusses two areas in which AI may have led cognitive science in inappropriate directions and suggests that a certain kind of theoretical and empirical developmental psychology can contribute to theory development in cognitive science. First, there are increasing doubts concerning the validity and usefulness of computational and cognitivist views of ability, which forces us to address the foundational question of what cognition is. The computational metaphor can be elaborated in more than one way, and AI has not necessarily stressed the most appropriate dimensions. Second, when we come to address the issue of cognitive change, the results of the field of machine learning—despite their initial promise—have proved disappointing.

DRAWING THE BOUNDARIES OF COGNITION AND COMPUTATION

Figure 1 shows Pylyshyn's scheme for partitioning the world into sources of behavioural complexity.[3] It does not aim to depict a mechanism for intelligent abilities, and its author insists that it is 'simple-minded'. Nevertheless, it inadvertently succeeds in illustrating two important character-

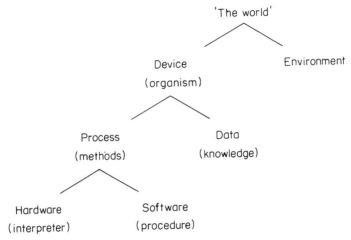

Fig. 1 Pylyshyn's scheme for possible sources of behavioural complexity

istics of mainstream AI models. One is that subject—device or organism—and environment are seen as alternative sources of complexity, with little or no attention paid to the environment. (Pylyshyn's commentary is rapidly dismissive of the ecological psychology proposal that a far greater share of complexity must be ascribed to the environment.) In keeping with the establishment of this internal–external dichotomy, behaviour does not appear at all. (Pylyshyn is not unique in taking 'behaviour' to mean something like 'what the subject can get done', directing attention towards task structure and away from observable body movements.) AI's version of the computational metaphor, cognitive psychology and cognitive science shares a tendency to view mind and brain as coextensive and cognition as something which occurs 'between the ears'. Within cognitive science, this is taken to an extreme in the 'methodological solipsism' advocated by the philosopher Fodor's influential version of a 'computational theory of mind'.[4] He argues that a science of cognition cannot be based on a computational system's connections to the world, in part because he sees no possibility of a 'naturalistic' psychology specifying lawful relations between mental states of an organism and objects in its environment. His conclusion is that mental processes can and should be studied as computations defined over formal properties of representations of the environment, without access to semantic properties such as reference or meaning.

The present paper does not support this perspective, but believes that getting the subject back into the environment is essential to defending a computational approach against mounting criticism. Outside AI, a nascent ecological psychology develops the theory of 'direct perception' (e.g.[5,6]). The metaphor of a radio receiver 'resonating' to environmental information is preferred to that of a computer 'processing' it, and the image of a passive computer is juxtaposed with that of an active subject who seeks out information rather than constructing it. While wishing to endorse ecological psychology's critique of the overly constructivist nature of mainstream AI and cognitive psychology, it will be suggested that computational notions retain a great deal of utility. Within AI, Winograd has developed some of the most interesting criticisms of computational approaches, originating with what he perceives to be the failure of his own and other AI research in the domain of natural language). Relevant here is his desire to break away from AI's 'rational' model of ability in which the subject is assumed to have an explicit uniform representation of a situation and rules for choosing operators which can transform it to reach a desired goal. For example, Flores and Winograd note that an AI program explaining a newborn baby's ability to get food would probably include: a set of 'goals', such as 'drink milk'; a set of 'operators', such as 'cry' and 'suck'; a model of the world setting them in correspondence; and possibly even a model of adult behaviour. Their criticism of AI's tendency to place everything believed necessary for the attainment of some outcome into the subject is appropriate, as will be shown below. Before heeding their caution that infants do not really 'know' anything—the actual mechanisms involved being simply reflexes—it is informative to consider psychological studies of human infants.

The nature and organization of infant abilities suggests how we are preadapted to interact with and develop in the world. Considering which emerging cognitive science concepts are of greatest utility in helping us to understand these abilities can be of reciprocal benefit to cognitive science by directing us to those aspects of the computational metaphor which map onto the foundations of mind. Here, it is argued that infant abilities lie in action, which is situated in the world and best conceptualized computationally. Three basic themes will be discussed in the following subsections.

Functional perceptual–behavioural coordination

Newborn and very young infants have been shown to exhibit an increasingly wide range of what can be termed 'functional perceptual–behavioural coordinations' with respect to objects and persons (cf.[9,10]). For example, coordinations found in newborn and young infants prior to relevant practical experience include: reaching and grasping for objects;[11–13] manual interception of moving objects;[14] 'defensive' behaviours with objects moving on a hit path towards the face;[15,16] 'imitation' of adult gestures such as tongue protrusion and hand movement;[17,18] the beginnings of 'communication' with persons in the form of recurrent patterns of face and body gestures, together with distinctive mouth and tongue movements;[19–22] and locomotion on a supporting surface[23]).

Close examination of such abilities and of their development supports the need for a type of computational model of the general form sketched in Fig. 2.[24] Space limitations preclude discussion of the derivation or details of this model, but a minimal account follows. The claim that the infant is

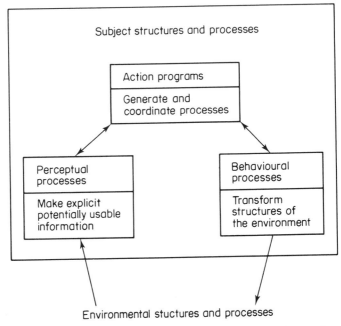

Fig. 2   Main components of a computational model of the infant action system

best understood as an 'action system' is based on treating action as an organizing construct. Perception and behaviour, subject and environment, and the relationship between the program and physical machine are all necessary to this model:

1. Environmental regularities make it possible for perceptual processes to construct symbol structures or descriptions making explicit information in the world. Different information is potentially relevant to different activities.
2. The selective use of information to generate behaviour is made possible by action programs which control the functional coordination of perception and behaviour.
3. Behavioural processes include those which manipulate environmental structure (e.g. reaching and grasping) and those which manipulate the relationship between subject and environment (e.g. head and eye movements). The latter make it possible for action programs to control which of the many potential input images in any context will actually be processed—significant for selective attention.

   Three main points are relevant to AI's use of the computational metaphor:

1. For several of the infant abilities mentioned above, the visual information involved is relatively well understood. The findings cast doubt

on AI approaches which stress that all physical events are inherently ambiguous. There is more support for Marr's view that there are relations, patterns or *constraints* in the way entities in the physical world interact with each other, and that perceptual processes exploit this fact in order to *recover* information about the properties of objects and events from sensory input.[25] Marr's assumption that the early levels of vision are data driven and do not rely on individual experience or memory implies that we could expect such processing in infants. In fact, the framework of different levels of processing or description which Marr develops proves very enlightening with respect to findings and controversies in the infant perception literature.

2. One attraction of AI is that its computational approach appears to provide an active metaphor for knowledge. It offers process-oriented accounts of ability which promise to avoid the reification and arbitrary boundaries found in other disciplines. For example, Minsky considers terms like 'know', 'believe' and 'mean' to be 'prescientific idea germs' which will be supplanted as we develop better concepts.[26] Our abilities are said to involve the (often complex) interaction of many processes, and to demand what Haugeland calls a '*systematic*' style of explanation[27]—one which models abilities in terms of organized interactions between components which cannot achieve their functional role independently of one another. Thus, it is disappointing to find that many psychologists see the computational metaphor as essentially passive (e.g.[28,29]).

The model of infant action supports the need for a process-oriented systematic explanation, but suggests that the boundaries of an action system need to be redrawn more broadly to take account of environmental and behavioural processes. The philosophical forms of functionalism to which AI and Fodor are committed focus on issues concerning mental states. As an alternative, it may prove fruitful to return to the psychological version of functionalism, with its emphasis on action, adaptation and the practical consequences of the mind in the real world (see Schultz[30] for an historical account of this school of psychology). For example, infant perception is more appropriately and actively viewed in terms of the selection and use of information to guide action than as the construction of a mental state such as a percept or perceptual belief.

3. From a developmental perspective, the *relationship* between a computer program and the physical machine in which it is implemented is just as interesting as the independence which is emphasized in much AI. Of particular significance is the way it offers us a dual explanation of a system's functioning.[31,32] The operation of the program level is susceptible to a *purposive* analysis, yet the processes it controls are those of a physical machine—computer or body—obeying *causal* laws. This is an invaluable way of conceptualizing how the mind can actually

get things done in the world, and it is useful to the infancy model in two ways.

First, its applicability justifies considering infant behaviour in terms of action—purposive transformation of the environment—rather than reflex. Even what might be considered a single behaviour, such as reaching or sucking, will show complex restructurings which cannot be explained if we describe them as reflexes—involuntary coordinations whose organization does not show development. Using the computational metaphor, we can consider the preadapted movements which are present at birth or appear through maturation as providing the causal process basis of behaviour; its purposive aspect is achieved through program control and different types of program structure can make possible different forms or degrees of purpose or intention. Certainly, infant action has few of the hallmarks of mature purpose, but it does not lack them entirely (Boden discusses relevant criteria in detail[31]). The earliest coordinations reveal a 'definite end point' and a type of 'prospective reference'. For example, Bruner notes that while an infant reaches for an object, the mouth may open and close before the reach is completed, as if in 'anticipation' of the act of sucking the captured object.[33,34] A finger placed in the mouth before the object is grasped will be sucked and the reach stopped, suggesting that 'The object is plainly destined for the mouth'.[33] Other features of purpose, such as flexibility or variation in the execution of behaviour, are initially absent. However, the changes in the execution and ordering of behaviours which mark their development can be considered to arise through a reorganization of the programs controlling them.

Second, applying this aspect of the metaphor to infant abilities highlights the fact that program control is of both internal and external transformational processes. Covert 'intellectual' processes manipulate internal symbol structures; overt behavioural or motor processes have a privileged status as they can manipulate external environmental structures. This unique view of behaviour implies that we need to take it more seriously. It is part of the process of knowing, not simply a product or 'output' offering evidence of some static internal knowledge.

This perspective proves important to understanding the role of action in extending our knowledge of the world, an issue which is taken up later.

Meaning and 'causal embedding'

We may see infants produce appropriate behaviour, but is there really meaning or understanding on the part of the infant? This problem corresponds to a controversial issue in cognitive science: under what conditions can a machine be said to refer to the world or attribute meaning to the symbols it manipulates? One solution to this is to argue that the machine would need to have sensory and effector apparatus and be *causally embedded* in the world to which its symbol structures refer. The model of infant ability, sketched above, converges in an interesting way with McDermott's philosophically motivated argument for this type of solution.[35]

As adults, McDermott suggests, we are misguided in our intuition that internal symbols have meaning for us because we interpret them, i.e. have some way of connecting them to concrete external reality which tells us what they refer to. We may never refer to the world in a manner different from a robot, which can manipulate symbols without knowing what they 'mean' in this way. A (purely formal) system of this type can work when three conditions are satisfied:

1. *Input soundness.* Symbol structures constructed by the sensors are usually 'true', i.e. correspond to actual states of affairs in the world. If a car were coming, (CAR-COMING) would be entered into the system.
2. *Inferential soundness.* The formal system is constructed so that it tends to infer 'true' symbol structures from others. On the basis of (CAR-COMING) it needs to infer (SHOULD-DO (JUMP)).
3. *Output soundness.* The effectors are constructed so that they tend to perform the behaviours assigned to the symbol structures connected with them. When the symbol structure (JUMP) is sent to the effectors, a jump is actually produced.

If we consider the explanation of infant abilities in terms of action, then all of these conditions are satisfied. Say an object is moving towards the infant's face. There is input soundness as 'looming'—symmetrical expansion of an optical contour in the middle of the visual field—is recovered from the input by the perceptual component. Action programs underlie inferential soundness; on the basis of this information, a program concerned with avoidance is activated. The ability of this program to selectively generate and control 'defensive' behavioural processes such as backwards head movement and arm raising demonstrates output soundness.

This description, which is used to make sense of the infant system, may technically be called a 'standard sound interpretation'. McDermott stresses that such an interpretation is a logical tool to explain how the system works; it is not something that the system itself needs or has access to over and above its symbol structures. Developmental and computational views that information from the world must be interpreted by matching against internal models or concepts are being challenged (e.g.[6,9]). The causal embedding approach to meaning offers a principled alternative within the computational metaphor. It clarifies how meaning can be implicit in action.

Externalization and scaffolding

Clearly, we cannot understand a computational system solely in terms of the construction and manipulation of intrasubject symbol structures. The abilities of naturally intelligent systems rely extensively on an externaliz-ation of processing which exploits and is scaffolded by physical and social environmental structure. Bruner coined the term '*scaffolding*' to describe the way adults help children to solve problems by reducing the number of degrees of freedom involved. A child who cannot yet walk may be supported by the arms and moves forwards, offering them the opportunity to make the necessary leg movements and to experience attaining the 'goal' of locomotion which they will eventually be capable of achieving alone.

Kaye's detailed studies of infant–adult interaction—from the first feed to language use—lead him to conclude that the possibility and organiz-ational sequence of many early abilities stems as much from the adult partner as from the infant.[36] He finds Goffman's notion of 'frames' useful to describe regular patterns of adult behaviour which provide the spatial and temporal structure in which the infant develops and displays increas-ingly sophisticated skills. It is interesting to note that AI has also developed the notion of 'frames' to capture a form of knowledge organization (e.g. see[37]). In AI, however, a frame is a type of data structure containing information about a stereotyped situation. It can be accessed from memory in new situations to classify them and work out what to expect and do.

Can these concerns be extended beyond infancy? We might concen-trate on a possible developmental transition between the two 'frame' notions considered above, posing the common question: how does external structure become internal? Here, a rather different direction will be taken, by emphasizing an essential continuity between infant, child and adult. One particularly relevant area of research will be discussed briefly. This is the cognitive psychology concerned with '*everyday cognition*'. Cole and his associates have undertaken many developmental as well as cross-cultural studies of intellectual functioning in 'real life' as opposed to more rarified laboratory settings. For example, Cole, Hood, and McDermott investigated remembering among 8 to 10 year old children who were engaged in cake-baking and seed-planting activities.[38] A major finding was that the physical structure of the environment was repeatedly used as a memory aid. When a recipe calls for melted shortening, it is sought on the shelves where ingredients are stored. Working out the number of seed trays needed for planting is facilitated by the presence of the fruit and vegetables which have just been cut up. As the authors note: 'The fact that many everyday environments (e.g., kitchens) are physically structured in a manner that lessens memory load is so ubiquitous that it is difficult to see (p. 368).'

Levin and Kareev's study of 10 year olds' problem solving at a computer club emphasizes the way these children draw on each other as a 'rich external resource'.[39] In one computer game, the problem is to direct a screen harpoon at a screen shark. A typical episode revealed a division of labour in which one child provided the 'bearing', a second provided the 'range' and a third acted as typist for these suggestions. An important point which emerges is that AI and experimental concerns—such as the way a problem space is represented and productive moves sought—*can* be related to the everyday situation. However, this is frequently a social situation, and a real-life theory of problem solving needs to include processes such as 'help seeking' and 'help using'.

Even in adulthood, it remains useful to get one's cognition into the environment, such as when we use paper and pencil or rely on a familiar supermarket layout. Although internal structure does change, this does not remove the subject from the environment but extends the subject's potential for action in the world. Developmentally speaking, we can view society as exploiting the preadapted organizational form of ability. Social and technological cultural tools serve to embed infant, child and adult in a network of culturally prescribed meanings.

HOW DOES DEVELOPMENT WORK?

In attempting to understand learning and development, a straightforward relationship between AI and developmental psychology suggests itself. Developmental psychology provides the observations of what children can do at different points in development, and AI offers theories of the mechanisms which could explain the observed changes. It should, therefore, simply be a question of bringing these two areas into alignment. Unfortunately, this is not the case, and both disciplines contribute to the difficulty. Most AI research is guided by ideas about learning which are plausible but may be misguided. Most developmental psychology provides the wrong type of data. As noted earlier the tendency is to concentrate on outcome measures of success or failure, not on the process by which a particular outcome might be attained. The following subsections consider mainstream AI's approach to learning and then go on to look at recent developmental research which suggests new directions of inquiry.

The AI view of learning

A representative example of an AI learning program is Sussman's HACKER,[40] which solves problems in a microworld of spatially arranged blocks. It has information about restrictions on its own behaviour (e.g. it can only pick up one block at a time) and about conditions in the world which must be met if particular types of behaviour are to be executed (e.g. that a block must have a CLEARTOP if it is to be picked up). HACKER is able to improve its skill in this domain through processes which locate, diagnose and rectify 'bugs' or points at which its programs fail to run smoothly. It would be inaccurate to say that the design of HACKER does not produce interesting insights. For example, one might assume that there are a vast number of possible bugs. On the contrary, at the level of HACKER's plans for getting from a start to a goal state, Sussman is able to show that types of bugs can be mapped onto a very restricted number of patterns of program structure. Most involve either unsatisfied preconditions or protection violations in which the execution of one behaviour would make execution of others impossible or eliminate the preconditions they require. However, two general problems stand in the way of adopting HACKER as a model of human learning.

First, the design of HACKER is based on Sussman's intuitions about problem solving and learning. Sussman notes that his anthropomorphism in referring to HACKER as 'he' reflects that the system's development was guided by introspection on his own problem-solving behaviour and not that he claims that the same mechanisms operate in people. Flores and Winograd[7] make the straightforward yet non-trivial point that a programmer can only program what they are conscious of. This provides a strong case for the relevance of empirical developmental research. In looking at different ages and domains of activity, we may become aware of regularities which escape our individual adult introspection.

Second, the scope of HACKER's learning is relatively restricted, focusing on plans and strategies. The system is predicated on a definition of learning in terms of processes which improve performance in reaching a goal—a common enough idea in AI. Indeed, Mitchell suggested that learning is made unnaturally difficult unless the subject knows what the goal is.[41] Thus, HACKER is given a series of problems and gets better and better at reaching the goals specified in them. But HACKER does not really construct novel goals for itself. Similarly, many of HACKER's reorganizations of its plans are possible because of its knowledge of preconditions for behaviour. But HACKER does not appear to discover new preconditions for itself, although it can learn to 'recognize' that in certain circumstances a particular type of strategy is needed for success. Neither of these restrictions are true of developing human systems.

Recently, Scott and Vogt have expressed reservations about AI learning research which converge with these developmental considerations.[42] They emphasize the need for knowledge-based AI systems which can acquire their own knowledge base, and suggest that machine learning has paid too exclusive attention to *task-oriented learning* (TOL). A complementary approach should deal with *knowledge-oriented* learning (KOL). This may be defined as 'the construction of an organized representation of experience'[42] and is not motivated by the attempt to improve performance on any particular task. Scott and Vogt suggest that the relationship between KOL and TOL may be like that between pure science and technology. In science, these two aspects are complementary, and they may need to be in a more successful machine learning. Pure science is concerned with extending our understanding of how the observable world works, but the results of its investigations may have long-term, unforeseen practical benefits. For example, nineteenth century investigations provided knowledge of electromagnetic phenomena which is fundamental to much of our current technology. Scott and Vogt are sensitive to the fact that the activities of young naturally intelligent systems may be viewed as a form of KOL. The following subsections discuss research which confirms this suggestion.

Microanalytic developmental studies: Taking behaviour seriously

Asking how naturally intelligent learning differs from most current machine learning establishes a close link with Winograd's and Flore's critique of mainstream views of computers and cognition.[7] They stress that a system cannot solve a problem by searching rationally through a problem space until that problem space has been created, but human intelligence is able to cope with situations which do not appear to be represented in this manner. Winograd's and Flore's alternative framework draws heavily on the philosophy of Heidegger. This may seem to be a digression, but it turns out to provide a further justification for developmental study.

Four concepts are central. In everyday life we appear to be 'thrown' into activity, being forced to act even when we are unable to reflect on a situation and select a course of action. Objects and properties of the world which are employed in the course of action are generally not explicitly recognized by the subject who appears to use them. They are 'ready-to-hand', part of the background which is taken for granted. Only when there is a 'breaking down' of activity do they become explicit or 'present-at-hand'. For example, gravity is ready-to-hand for the infant learning to walk or handle objects; it becomes present-at-hand for someone designing a space vehicle as they must deal with breaking downs which result from its absence.

This account bears a striking resemblance to Piaget's theory of the development of knowledge, which has provided such a significant focus for developmental psychology.[43–45] He argues for a distinction between two autonomous forms of knowledge: underlying action and conceptualization respectively. We repeatedly achieve *practical success* prior to any *theoretical understanding* of how or why our actions succeed. Disturbances provide the motivation for equilibration processes, which are responsible for the developmental transition between success and understanding. This involves a progressive conceptualization of action, associated with developing cognizance or a 'grasp of consciousness'. Consciousness of the goal achieved precedes consciousness of means. For example, Piaget[43] demonstrates that children who can use a sling-shot to get a ball into a box do so with little or no grasp of what they are doing. Initially, they are convinced that they let the ball go when it is directly in front of them—the box's location—and not at their side, from which it actually follows a tangential path to the box.

This correspondence between viewpoints is not necessarily a good thing. The framework which Winograd and Flore's offer as a solution to the difficulties faced by AI models actually marks major and long-standing problems in developmental psychology. Piaget failed to clarify how the equilibration process could work, and increasingly it seems that new concepts are needed. The most promising alternative is offered by what will be called 'microanalytic' developmental methods. The earlier discussion of behaviour stressed that it should be treated more seriously in attempts to understand the mind. This is precisely what such methods do, concentrating on the fine detail of behaviour and how that detail is modified as the subject repeatedly functions in some context. The most interesting models being developed are explicitly informed by computational ideas, although their perspective on processes of change differs from that of AI. It proves particularly instructive to compare change within different age periods, so examples will be offered from work on infant grasping[46] and block balancing in the middle years of childhood.[47]

Mounoud and Hauert record the ('cinematic') details of grasping in 6 to 16 month old infants. Among 6 to 8 month olds, there is a functionally adequate pattern of behaviour—involving jointed contractions of antagonistic muscles—which permits effective grasping (without over- or undershooting of the arm) whatever an object's specific weight. This type of behaviour is equally effective if, for instance, an object has been grasped several times before it is replaced by a much lighter one of identical size. In contrast with this, the grasping of older infants shows disruption, such as sudden upward movement of the arm in the face of this substitution. By 14 to 16 months of age, this excessive arm movement also occurs in the substitution context, but now it is rapidly compensated for. Between 9 and 16 months, an action-based understanding of size–weight covariation appears to have been established. The outcomes of perceptual and proprioceptive functioning in this situation lead to reorganization of the programs controlling the relevant movements. The way in which the hand is contracted *prior* to grasping becomes specific to object weight, which is *inferred* from perceived size.

Karmiloff-Smith and Inhelder presented 4 to 9 year old subjects with blocks which were either of uniform density, or visibly or invisibly weighted at one end.[47] Even the youngest child was able to balance any of the blocks, apparently making use of proprioception in order to move them back and forth along the bar until their balance point was found. With further experience, however, they tried to balance every block at its geometric centre, including the weighted blocks. These were put to one side as 'impossible' when they failed to balance. Beyond this, a final level of performance appeared in which the geometric centre was tried first, but failure with weighted blocks now led to a readoption of the proprioceptive strategy of the first level. The child may now impose an order on which type of block is tackled. If one unevenly weighted block is balanced, the child will look for a similar block to try. Initially, blocks did not tend to be handled in any particular sequence.

Phases of development

Several conclusions emerge from comparing these and similar studies. The most central underlines the limitations of notions such as failure, disturbance or breaking down as motivating forces for development. Subjects begin by being able to grasp or balance an object, so why should they change? Karmiloff-Smith for example,[48,49] suggests that change is *not* error driven; the subject's motivation is greater control over both the environment and his or her own processing. If one simply considered the outcome of performance from an observer's viewpoint, two periods of successful activity appear to be separated by one of disruption or error, but this is evidence of a reorganization which is not a regression from the perspective of the subject's representational system. Both Mounoud and Hauert[46] and Karmiloff-Smith[49] believe that computational ideas are necessary to explain such phenomena. Their perspectives are by no means identical, but they do see behaviour as a program-controlled process and consider reorganizations in these programs to explain changes in their subject representation of the world.

A second general conclusion is that change involves making explicit the constraints on successful action. Say, for example, an infant pulls a green blanket and discovers that this brings an attractive toy within reach. The behaviour succeeds for any object which IS-ON any other object within reach, but this invariant relationship is embedded in a vast amount of situation-specific information about the world and the infant's own behaviour. How is the infant to know that this action was successful not because of the blanket colour or shape of the toy but because the toy was on the blanket? In the studies discussed above, the infant comes to take object size into account before grasping; the child eventually anticipates that blocks should be balanced on their centre point. Karmiloff-Smith argues for a general three-phase model of development which addresses this in terms of implicit and explicit knowledge.[49] Her model draws on the computational metaphor, emphasizing the role and organization of program procedures for controlling processing.

The first phase is called *procedural* as it reveals the functioning of effective but, as yet, unrelated procedures. The phase-one child could be considered to balance a series of blocks or grasp a series of objects, without their procedural organization linking these 'individual' problems. Control of behaviour is primarily 'data driven', based on effects in the current situation. The second phase involves what Karmiloff-Smith terms *metaprocedural* activity. The procedures of phase one themselves become the focus of special procedures which are specifically concerned with organization. Their role is to mark similarities and differences among the successfully functioning procedures concerned, rendering explicit knowledge which is only implicit in the aggregate of functioning of the phase-one level. This process gives rise to the 'theory-driven' behaviour of this phase. The child cannot cope with 'counterexamples' such as objects whose weight is lighter than their size predicts or blocks which do not balance at the centre. (Interestingly, children in this phase can balance what they call the 'impossible' blocks if they perform the task with eyes closed.) The final phase is termed *conceptual* by virtue of the flexible interaction which it reveals between the two earlier forms of behavioural control.

This discussion emphasizes similarities in processes of change at very different age periods and in distinct domains of activity. Such similarities are not due to the fact that both of the example studies concern sensory–motor coordinations. Karmiloff-Smith[48] presents ample evidence that her general model applies equally well to language behaviour. Thus, clear support is offered for Scott and Vogt's view of the wide-ranging significance of knowledge-oriented learning. In the developmental examples, children do not learn because they fail on the tasks involved; they actually construct a deeper understanding of what the tasks involve because they have succeeded in solving them.

CONCLUSION

It has been suggested that developmental psychology can make an important contribution to cognitive science. The computational metaphor is central to cognitive science, but the way in which AI has used it has resulted in increasing doubts about its relevance to explanations of human abilities. Developmental psychology may help overcome these doubts by showing how computational ideas are relevant to understanding naturally intelligent systems. The analysis presented in this paper leads to two major suggestions for future AI models. One arises from the theoretical analysis of infant abilities. Human intelligence cannot be understood solely in terms of internal structures and processes. From infancy, it is organized in terms of action and relies on exchange and transaction with a structured physical and social world. The other has its origins in developmental methods which take behaviour seriously. These suggest that plausible intuitions about how intelligence develops are deficient when compared with insights gained from the comparison of change at different age periods and in different domains of activity. Future interdisciplinary attempts to understand learning should find it profitable to bring together AI modelling techniques with the type of data offered by microanalytic developmental studies.

REFERENCES

1. Miller, L. (1978). 'Has artificial intelligence contributed to our understanding of the human mind? A critique of arguments for and against'. *Cognitive Science*, **2**, 111–127.
2. Johnson-Laird, P. N. (1982). *Mental Models: Towards a Cognitive Science of Language. Inference and Consciousness*, Cambridge University Press, Cambridge.
3. Pylyshyn, Z. W. (1981). 'Complexity and the study of natural and artificial intelligence'. In *Mind Design* (Ed. J. Haugeland), Bradford, Cambridge, Mass.
4. Fodor, J. A. (1980). 'Methodological solipsism considered as a research strategy in cognitive science'. *Behavioural and Brain Sciences*, **3**(1).

5. Michaels, C., and Carello, C. (1981). *Direct Perception*, Prentice-Hall, New York.
6. Turvey, M. T., Shaw, R. E., Reed, E. S., and Mace, W. M. (1981). 'Ecological laws of perceiving and acting: in reply to Fodor and Pylyshyn'. *Cognition*, **9**, 237–304.
7. Winograd, T. and Flores, C.F., (1985). *Understanding Computers and Cognition: A New Foundation for Design*, Ablex.
8. Winograd, T. (1980). 'What does it mean to understand language?' *Cognitive Science*, **4**, 209–242.
9. Butterworth, G. E. (1981). 'Object permanence and identity in Piaget's theory of infant cognition'. In *Infancy and Epistemology: An Evaluation of Piaget's Theory* (Ed. G. E. Butterworth), Harvester Press, Brighton.
10. Mounoud, P., and Vinter, A. (1981). 'Representation and sensorimotor development'. In *Infancy and Epistemology: An Evaluation of Piaget's Theory* (Ed. G. E. Butterworth), Harvester Press, Brighton.
11. Bower, T. G. R. (1972). 'Object perception in infants'. *Perception*, **1**, 15–30.
12. Bower, T. G. R., Broughton, J. M., and Moore, M. K. (1970). 'The coordination of visual and tactual input in infants'. *Perception and Psychophysics*, **8**, 51–53.
13. Bower, T. G. R., Broughton, J. M., and Moore, M. K. (1970). 'Demonstration of intention in the reaching behaviour of neonate humans'. *Nature*, **228**, 679–681.
14. Hofsten, C. von (1980). 'Predictive reaching for moving objects by human infants'. *Journal of Experimental Child Psychology*, **30**, 369–382.
15. Ball, W., and Tronick, E. (1971). 'Infant responses to impending collision: optical and real'. *Science*, **171**, 818–820.
16. Ball, W., and Vurpillot, E. (1976). 'La perception du mouvement en profoundeur chez le nourrisson'. *Annee Psychologique*, **76**, 383–400.
17. Maratos, O. (1982). 'Trends in the development of early imitation in infancy'. In *Regressions in Development: Basic Phenomena and Theories* (Ed. T. G. Bever), Lawrence Erlbaum, Hillsdale, N.J.
18. Meltzoff, A. N. (1981). 'Imitation, inter-modal coordination and representation in early infancy'. In *Infancy and Epistemology: An Evaluation of Piaget's Theory* (Ed. G. E. Butterworth), Harvester Press, Brighton.
19. Brazelton, T. B., Koslowski, B., and Main, M. (1974). 'The origin of reciprocity: the early mother–infant interaction'. In *The Effect of the Infant on Its Caregiver* (Eds M. Lewis and L. A. Rosenblum), Wiley, New York.
20. Trevarthen, C. (1974). 'Conversations with a two-month old'. *New Scientist*, **62**, 230–235.
21. Trevarthen, C. (1975). 'Early attempts at speech'. In *Child Alive* (Ed. R. Lewin), Temple-Smith, London.
22. Trevarthen, C. (1984). 'How control of movement develops'. In *Human Motor Actions: Bernstein Reassessed* (Ed. H. T. A. Whiting), North Holland, Amsterdam.
23. Rader, N., Bausano, M., and Richards, M. E. (1980). 'On the nature of the visual cliff avoidance response in human infants'. *Child Development*, **51**, 61–68.
24. Rutkowska, J. C. (1984). 'Explaining infant perception: insights from artificial intelligence?' Cognitive Science Research Paper, Serial No. 005, University of Sussex, Brighton.
25. Marr, D. (1982). *Vision*, Freeman, San Francisco.
26. Minsky, M. (1980). 'Decentralized minds'. *Behavioural and Brain Sciences*, **3**(3).
27. Haugeland, J. (1978). 'The nature and plausibility of cognitivism'. *Behavioral and Brain Sciences*, **1**, 215–260.

28. Brown, A. (1979). 'Theories of memory and the problems of development'. In *Levels of Processing in Human Memory* (Eds L. S. Cermak and F. I. M. Craik), Lawrence Erlbaum, Hillsdale, N.J.
29. Neisser, U. (1976). 'General, academic and artificial intelligence'. In *Intelligence* (Ed. L. Resnick), Lawrence Erlbaum, Hillsdale, N.J.
30. Schultz, D. (1981). *A History of Modern Psychology*, 3rd ed., Academic Press, New York and London.
31. Boden, M. A. (1972). *Purposive Explanation in Psychology*, Harvard University Press, Cambridge, Mass.
32. Boden, M. A. (1981). *Minds and Mechanisms: Philosophical Psychology and Computational Models*, Harvester Press, Brighton.
33. Bruner, J. S. (1968). *Processes in Cognitive Growth: Infancy*, Clark University Press with Barre Publishers.
34. Bruner, J. S. (1973). 'Organization of early skilled action'. *Child Development*, **44**, 1–11.
35. McDermott, D. (1983). 'Under what conditions can a machine attribute meanings to symbols?' *Proceedings Eighth International Joint Conference on Artificial Intelligence*, Karlsruhe, Germany, Vol. 1, pp. 45–46.
36. Kaye, K. (1982). *The Mental and Social Life of Babies*. Harvester Press, Brighton.
37. Minsky, M. (1975). 'Frame-system theory'. Reprinted in *Thinking: Readings in Cognitive Science* (Eds P. N. Johnson-Laird and P. C. Wason), 1977, Cambridge University Press, Cambridge.
38. Cole, M., Hood, L., and McDermott, R. (1978). 'Ecological niche picking'. Reprinted in *Memory Observed: Remembering in Natural Contexts* (Ed. U. Neisser), 1982, Freeman, San Francisco.
39. Levin, J. A., and Kareev, Y. (1980). 'Problem solving in everyday situations'. *The Quarterly Newsletter of the Laboratory of Comparative Human Cognition*, **2**, 47–52.
40. Sussman, G. J. (1975). *A Computer Model of Skill Acquisition*, American Elsevier, New York.
41. Mitchell, T. M. (1983). 'Learning and problem-solving. *Proceedings Eighth International Joint Conference on Artificial Intelligence*, Karlsruhe, Germany, Vol. 2, pp. 1139–1151.
42. Scott, P. D., and Vogt, R. C. (1983). Knowledge oriented learning'. *Proceedings Eighth International Joint Conference on Artificial Intelligence*, Karlsruhe, Germany, Vol. 1, pp. 432–435.
41. Mitchell T. M. (1983). 'Learning and problem-solving. *Proceedings Eighth International Joint Conference on Artificial Intelligence*, Karlsruhe, Germany, Vol. 2, pp. 1139–1151.
42 Scott, P. D., and Vogt, R. C. (1983). 'Knowledge oriented learning'. *Proceedings Eighth International Joint Conference on Artificial Intelligence*, Karlsruhe, Germany, Vol. 1, pp. 432–435.
43. Piaget, J. (1976). *The Grasp of Consciousness*, Routledge and Kegan Paul, London.
44. Piaget, J. (1978). *The Development of Thought: Equilibration of Cognitive Structures*, Basil Blackwell, Oxford.
45. Piaget, J. (1978). *Success and Understanding*, Routledge and Kegan Paul, London.
46. Mounoud, P., and Hauert, C. A. (1982). 'Development of sensorimotor organization in young children: grasping and lifting objects'. In *Action and Thought: From Sensorimotor Schemes to Thought Operations* (Ed. G. E. Forman), Academic Press, New York.
47. Karmiloff-Smith, A., and Inhelder, B. (1974). 'If you want to get ahead, get a theory'. *Cognition*, **3**, 195–212.

48. Karmiloff-Smith, A. (1979). 'Micro- and macrodevelopmental changes in language acquisition and other representational systems'. *Cognitive Science*, **3**, 91–118.
49. Karmiloff-Smith, A. (1984). 'Children's problem solving'. In *Advances in Developmental Psychology* (Eds M. E. Lamb, A. L. Brown and B. Rogoff), Vol. 3, Lawrence Erlbaum, Hillsdale, N.J.

PART 3

AI—Culture and the Arts

Artificial Intelligence for Society
Edited by K. S. Gill
© 1986 John Wiley & Sons Ltd

9.   SHOULD ARTIFICIAL INTELLIGENCE TAKE CULTURE INTO CONSIDERATION?

PARTHA MITTER School of African and Asian Studies, University of
Sussex, Brighton

When I was invited to take part in the proceedings of the Conference on
Artificial Intelligence for Society, organized by the SEAKE Centre of the
Brighton Polytechnic, I felt both elated and apprehensive. As a historian
concerned with the humanities, I had only a lay person's untutored fascination for the technological miracle which artificial intelligence has
brought about, particularly the fundamental changes in our approach to
knowledge and the general processing of information. And yet, the very
theme of the Conference, which investigates the implications of the new
technology for society, especially the transfer of knowledge from the First
to the Third World, had a strong attraction for me. Increasingly, the new
technology is seen, perhaps rightly, as a powerful tool that could be
universally employed in social engineering to improve the lot of mankind.
Behind this optimism that the technological revolution will be able to
improve the quality of life in a society, or shall we say societies, rests the
implicit assumption that the needs of all societies are the same. The
assumption itself originates in the deep-seated empiricist epistemology of
the West that universal rules can be drawn up regarding human behaviour,

parallel to the behaviour observed by science in the natural world. Following from this, the votaries of artificial intelligence express confidence in their ability to transfer the latest European technology to non-European societies for the purposes of social engineering, irrespective of wide cultural differences.

Behind the confidence that knowledge can be transferred without taking into account cultural diversity lies the specific theory of knowledge which I hope to clarify by taking the example of AI. Computers, which are concerned with the acquisition and storing of knowledge, naturally take the human brain as their model, mimicking its deductive and memory functions. Computers are simply seen as faster and more efficient versions of our thinking process. How does a computer work? The hardware part of it executes the job that the programme or software instructs it to perform. Both hardware and software parts of cybernetics are the products of and governed by Western empiricist epistemology. In fact computers are the most recent in the succession of triumphs of Western science since the eighteenth century, based as it is on observation and experiment. Indeed the whole modern world is sustained by the foundations built by European science and technology, and it cannot for a moment be denied that within what Kuhn calls 'the paradigms set by Western science' the results are impressive and its precise strength lies in its ability to formulate laws that have universal validity.

Of course both the hardware as well as programming rely on empiricism, but the hardware or machine side of computers does not pose as much problem with such epistemology as the software. Programming which sets taks for the computing machine includes the domain of human needs. The burden of my argument in this paper will be to underline the point that there is no universality in human needs. The application of a universal scientific rationality to the workings of society, which even many social scientists endorse, ignores the immense diversity of human needs both within a society and even more so across cultures. To give an example from natural science: every time you dip an iron nail in copper sulphate solution, it will become copper plated. This experiment can be repeated *as infinitum* with the same outcome. In the case of human beings such uniformity does not exist. There is indeed a crucial difference between human and robotic thinking processes. Both use deduction and memory in their thinking process but human thought contains an extra element grounded in human experience and culturally conditioned. This is the realm of values and does not follow the universal laws of deductive logic. There is a lot of talk of interaction between computers and human beings. If we say that the computer emits a signal, how will it be interpreted? Its interpretation by a human being will depend on his/her cultural conditioning. The anthropologist Clifford Geertz[1] gives us the amusing example of the 'wink', which in a culture not conversant with this gesture may be taken to be a muscular twitch. The reason why we need to consider cross-cultural perception is because if knowledge transfer is to be meaningful it cannot ignore the cultural dimension.

What is culture? Here I shall refer to Popper's definition that will be useful here. The acquisition of knowledge is an activity that is not confined to human beings but is shared with the animal world and, with the advent of AI, with robots as well. Popper[2] suggests that knowledge can be simply viewed as problem solving, by forming hypotheses about the external world. The world of the living evolves setting aside those hypotheses that had been proven to be false and moving on to fresh ones. We human beings are no exception to this.

There is something unique to human beings, however. Unlike animals, knowledge in human society is accumulated and stored independently of the existence of individual human beings. Popper's example here is illuminating: when for instance someone writes a book, it is only natural that all sorts of personal and subjective elements will enter into it. However, once the book is finished and published, the book leads an autonomous existence quite independent of its author. It makes a contribution to the sum total of human culture. This example is offered here to underline the fact that it is culture, or products of the human mind, that makes human experience so unique. The importance of the specific cultural experience is often forgotten or ignored in scientific discussions, partly because, after the great advances made in physical sciences during the nineteenth century, the notion of 'pure' objectivity came to take a firm hold on the intellectual imagination. There can be no doubt about the impressive achievements of science in the field of the natural phenomena following on from the objective scientific method developed through the employment of observation and experiment. The scientific method's ability to predict and formulate general laws enabled it to harness the vast untapped forces of nature. The latest information technology is yet another triumphal demonstration, if such a demonstration was needed, of such a method.

We have seen how in the case of the physical phenomena the great virtue of the scientific method is its universal application. Certain problems surface when the same principles are applied to the realms of human experience, to the products of the human mind, in other words, to cultural systems. A persistent and pervasive assertion of the social sciences in the nineteenth century was that scientific objectivity transcended cultural varieties. An inevitable conclusion drawn was that all human beings behaved and thought in more or less the same way. This tendency, which can be called scientific reductionism, was, for instance, powerfully endorsed by Spencer and other Social Darwinists. Because it conceived the history of human civilization as a unilinear movement culminating in the modern European society of the nineteenth century, it viewed all non-Western cultures as poorer versions of Western civilization. Open any book on social and political thought and institutions written during the Victorian era and you will be struck by the almost unanimous representations of the so-called primitive societies as the childhood of mankind, whose maturity is reached in nineteenth-century Europe. There is a widespread failure to recognize that these societies were not the elementary versions of European society but may have developed their institutions independently. So it mattered little whether their progress had any connections, direct or otherwise, with developments in the West.

These persuasive doctrines of cultural homogeneity of the human society, which evolved in the nineteenth century and went in tandem with a hierarchical classification of cultures, even now informs a great deal of scientific thinking and frequently determines future development strategies of AI. This is the kind of thinking that is behind the optimistic assertion that the new technology will bring immeasurable benefits to countless millions in the Third World countries, irrespective of their specific cultural requirements. There is, however, a growing debate which questions this optimism generated in the formative period of the growth of information technology, and we have not yet gone beyond it. The CAAAT project of the SEAKE Centre belongs to the 'Luddites' that rightly reject such an undifferentiated picture of human society and stresses the need of the technological revolution to take into account both the variety and complexity of human cultural experience. The point at issue is not that there are no common grounds among human beings who are each one of them 'an island', so to speak. There are of course common and universal traits that we human beings share, but it is equally important to remember that there is also a great deal that is culture-specific. In short, the cultural factor cannot be discarded in the domain that deals with the acquisition and the processing of knowledge.

KNOWLEDGE AND CULTURE

Why should we be concerned with the role of culture in the field of AI? Here we need to go back to the distinction between the machine part of the computer and the programming for it. The machine does all that the programme commands it to accomplish—no more and no less—and that programme itself is a product of Western epistemology which in its turn is naturally influenced by the intellectual tradition or the prevailing values of the West. Yet how often is this realized when we speak of the effects of the new technology on mankind? I wish to go even further and suggest that Western epistemology itself is determined by the way the West perceives other societies. Recent works on the perceptions of other cultures have given us the insight that not only our perceptions of alien societies are conditioned by the manner in which we defined ourselves but our own self-definition is in its turn influenced by our perceptions of other cultures.

My task here is to bring out some of the pitfalls inherent in the unquestioned acceptance of certain doctrines that had gained currency during the previous century to which we are still heir, the doctrines which effectively ignored the important function of culture in matters pertaining to human affairs. Here I would turn to my own work on European representations of Indian society.[3] The work aims to bring out in all clarity what happens when two societies that have very different and even antithetical canons of culture come face to face with one another. Here the example of Western representations of another society is chosen but the example is meant to reflect a universal behaviour pattern and is not confined to the West. The general nature of how we perceive other cultures would be the valuable lesson for us by a close analysis of European representations of alien traditions.

A clear picture of European representations of the alien emerges in early travellers' reports and illustrations provided by them. The stereotype is the most common and frequent device for representing other cultures in this period which spans from the fourteenth to the eighteenth centuries. One of the meanings of the stereotype in the dictionary is 'a static image', which underlies the popular view of stereotypes as general, prejudiced impressions of other social and cultural groups. Perhaps there is no more vivid an example of this than the European commonplace that all Orientals look alike. This inability to distinguish individual features is not confined to Europeans. I knew someone in India who, because of his inexperience, could not distinguish between a pretty European girl and a plain one. To him they all looked alike.

If the simplified image is the distinct feature of the stereotype, then nowhere is there a greater scope for stereotyping than in description of alien unfamiliar societies. Victorian literature is full of stories of how Africans were simultaneously indolent and childish, sly and ferocious. Curtin's classic work has shown that these impressions were not based on actual observations of individuals.[4] From this description we need not jump to the conclusion that stereotypes are a modern phenomenon. After all, that venerable Roman authority on architecture, Vitruvius, after commenting on the effect of climate on different races, came to the conclusion: 'Heat quickens southern intelligence, inasmuch as it slows down the northern European, making him sluggish. Likewise manliness is sucked out by heat, making southerners cowardly'.[5]

It is held with some justification that stereotypes offer ingredients for prejudice against other cultural groups, as shown in Allport's pioneering study.[6] However, not all stereotypes are necessarily hostile. An instance of this is the commonplace among literate Europeans that Indians are spiritual. What is the psychological process behind stereotyping? An experiment conducted by Allport will help to explain it.

In a controlled experiment, a group of white Americans were first shown on the screen a New York subway compartment containing all white passengers except one black man among them. The next image flashed on the screen was of a holdup of the same compartment by a man wielding a knife. This particular image was flashed on the screen only for a brief instant. The man in question was one of the whites. Afterwards when the subjects of the experiment were asked to identify the man with the knife, they unanimously saw the black man holding the knife even if this was not the case. Why did they do that?

The crucial point is that the subjects of the experiment were not allowed to reflect but to reach an instantaneous conclusion. There is a strong suggestion here that in situations like this the brain cannot engage in what psychologists see as sequential, step-by-step 'logical' reasoning, but is forced to reach a swift decision based on fragmentary information. Stereotypes are precisely of this nature and are more like Platonic types or paradigms which the human brain grasps all at once as it is fed a whole matrix of diverse but relevant data. We universally apply such a cognitive tool whenever we experience a new phenomenon, whether this has to do with concepts or with objects.

Why do we make use of simplified images to perceive the visual world? As Hochberg suggests, this had to do with our limited capacity to store information in our memory.[7] As a television ad has it: like human beings, computers have memory; unlike humans, the computer memory shrinks every time it performs a task. Though there is a world of difference between computers and the human brain, the metaphor does tell us something significant. Apart from a few random items, we are unable to retain data in our memory unless the data are encoded according to an ordering principle. In sum, 'classification' guides perception, which in its turn is triggered by 'expectation'. The Allport experiment makes it clear that white subjects 'expected' the knife to be held by the black man, according to what they had been led to believe.

Let us return to early Western perceptions of Indian culture. Of all aspects of Indian culture, the representations of Hindu gods with their many arms were the most intriguing to early travellers. One of the most celebrated illustrations of a Hindu god was the so-called deumo or the devil of Calicut. This image which has closer affinities with mediaeval European images of the devil than a piece of Indian sculpture, was executed by a German artist for the Italian traveller Varthema's travel account. As Varthema explained, the god in his book, this devil of Calicut, was worshipped by south Indians. In fact, here this composite image combines different representations of the devil in mediaeval European demonology. Yet whenever an Indian god needed to be represented in early European literature this composite monster stereotype stood for all Indian gods.

Perception psychologists explain to us why we tend to represent other cultures in this manner. In a famous example the noted art historian. Gombrich, demonstrates this problem:[8] when the medieval illustrator, Wolgemut, was asked to depict such different cities as Damascus, Constantinople and Ferrara, he adapted for his purposes the cliche of his own city of Nuremberg which he knew. All he did was to add a caption to indicate these exotic cities. Likewise, Varthema selected from his 'drawer of mental stereotypes' the devil image to represent the Hindu deity of Calicut. His mental set was guided by his cultural experience of the manner in which non-Christian religions were represented at the time.

The stereotype may be described as the preexisting schema acting upon the perceived object in a perceptual synthesis. In other words, our cognitive leap from the known to the unknown is mediated by means of these paradigms, or cognitive maps if you like. This is because the human brain cannot make sense of the bewildering array that is nature without imposing some kind of order.

The psychologist Oatley describes the mental set as 'the structure of representation within, of aspects of the world without'—a metaphor for the external world.[9] About a half century ago, Bartlett[10] had shown that mental sets or schemata were culturally determined. We actively organize our past experiences according to principles culled indirectly from our experience of the society we belong to. Bartlett's insight has important implications for 'cultural perception', where our inherited values constitute the requisite paradigms. Thus the stereotypes which are dependent on cultural preconceptions are a way of making sense of the alien and the unfamiliar.

If all this sounds too pessimistic, it is also true that stereotypes are not necessarily unchanging or permanent. Because our thinking process is a dynamic one, stereotypes tend to fade when factual, objective information overtakes preconceptions. The conclusion is that we carry our 'frames' of reference in our brains which are constantly modified under the impact of new information that arises out of our dealings with the external world. To repeat, therefore, stereotypes are the cognitive devices for making sense of the alien.

Stereotypes gain a firm hold on the human mind in the absence of adequate information. In early European encounters with India, the lack of information about the Hindu religion gave rise to monster stereotypes which predominated in texts and pictures. This situation changed substantially when in the nineteenth century there was a vast expansion of European knowledge about non-European societies, as explorer after explorer penetrated the remotest corners of the globe in order to record what they considered to be the most esoteric or bizarre customs. With this impressive growth in knowledge we could expect to find a greater understanding of other cultures. If in this we are disappointed, and we are disappointed, it would be fascinating to ponder on the reasons for this failure.

To put it briefly, the reason for the failure lay embedded in the nature of Western epistemology itself, with its formulation of objective laws based on observation that would have universal validity. The leading thinkers of the last century, and even our own, implicitly accepted that there existed scientific and objective rules about culture which only needed to be discovered and codified. The Victorians were most scrupulous about their 'dispassionate' objectivity, but they failed to recognize its limitations in that it was based on the assumptions of a particular tradition, in this case, the Western tradition.

LIMITS OF THE SCIENTIFIC METHOD

Secondly, this age accorded a special status to the notion of understanding. As already mentioned, under the impact of the spectacular avances in the natural sciences, scientific method came to be accepted as the sole respectable form of knowledge, and science was expected to provide us with an understanding of the world we inhabit. An epistemic legacy of this tradition is the tendency in popular parlance to make no clear distinction between 'knowing' and 'understanding'. In the physical sciences, the precise meaning of understanding is explaining, in other words, discovering causal relationships. The humanities fell under the spell of scientific induction, and cultural systems were subjected to its acid test, as it were. In the sphere of culture, however, where external workings of nature are not relevant but have to do with the inner workings of the mind, will this do?

The German historian and philosopher, Dilthey, was the first thinker who confronted this problem between external and internal workings of the mind, between the 'hardware' and the 'software' functions of the brain, to put it in the computer language of today.[11] Dilthey followed up this distinction and separated the natural sciences from the realm of human culture. As he pointed out, the difference between the arts and the sciences lay not so much in their content as their mode of analysis. Dilthey's importance lies in his clear demarcation of the domain of culture that cannot be subjected to scientific reductionism without violating its essential nature.

RELATIVISM AND UNDERSTANDING OTHER CULTURES

This insight gained at the turn of the century began to filter down first of all to those who were studying human behaviour, especially anthropologists because they were constantly faced with the problem of making sense of alien cultural systems. Translation of alien customs into the investigator's own cultural system began to emerge as an important anthropological device. The cultural distance between the investigator and the society investigated was seen as parallel to the problem of linguistic translation. Even a simple description of an alien culture may involve the translation of a whole world of ideas, values and actions into the researcher's own language.

The sympathy of the anthropologists studying other cultures for this view arose out of their rejection of the ethnocentrism of their intellectual forebears. The consequence was the growth of an extreme form of relativism which refused to apply any judgements of value. From now on, the anthropologists concentrated on studying the 'function' which each cultural manifestation had in the society that had given rise to it. They followed Wittgenstein's extreme relativism[12]. According to the Austrian philosopher, because, for instance, the rational scientist and the religious believer followed what he called two different 'forms of life', their experiences could never be compared or contrasted. To put it simply, the one could not understand the meaning generated by the other's statements.

The objection that might be raised against Wittgenstein is this: surely the scientist and the believer can grasp the meaning of each other's statements though the 'significance' of such statements may elude them. A further objection to the pessimism of the relativists that alien meaning can never be recovered can be raised. Translation of alien concepts may be difficult, which is not to say it is impossible. Bound by our own world views, we may not be able to respond fully, but the challenge can be met.

The significance of cultural relativism lies in demonstrating the drawbacks of judging other societies solely on the basis of one's own cultural values. The question is: is it possible to discard inherited values altogether when judging other cultures? There can be no doubt that in matters relating to culture, judgements of value are essential. To sacrifice judgement is to deny the importance of meaning, and meaning is central to understanding. As Gellner[13] holds, to regard everything within a culture as equally significant is self-defeating and does not lead to a better understanding. As human beings, we do understand other human beings on a certain level. It is also equally true that the uniqueness of a cultural system is anchored in a number of context-dependent cases of rationality. We may also raise the question, why are values so crucial for understanding? Without values to guide us in uncharted waters, we are not in a position to discover what standards or values govern others, let alone appreciate them.

The final question is: if we are thus pulled in two opposing directions by relativism and value judgements, is it at all possible to recover the meanings generated by alien cultural systems? This is possible, according to Dilthey, because of what he calls the objective nature of cultural expressions, reflecting the collective 'mentality' of a society, which forms the essential bridge between the perceiver and the perceived.[14] The required acculteration process begins in childhood when the individual child learns to interact with the environment; this enables it to place its subjective experience in a common context and into a shared world view.

Dilthey's insight came from his study of history, for he saw the otherness of the past as having parallels with cultural otherness. He found that our cultural experience consisted also of our collective memory of the past. Because history mediates between the present and the otherness of the past, reading a text from the past of one's own society involves retrieving alien meaning from a culture that was both similar and different. Thus culture makes us aware of the universality of the 'structure' of mental experiences, though their contents differ. We as interpreters of an alien culture are looking for equivalents in our own experience when we seek to make sense of an alien concept. The link between the interpreter and past to be interpreted, between the 'self' and the 'other', is the reconstructive or 'historical imagination'.

The empathy which is provided by 'historical imagination' is the empathy that aids in understanding an alien cultural system by releasing the investigator from the limitations of his or her own world view, on the one hand, and the 'willing suspension of judgement', on the other. To conclude, the potentials of the new cybernetic technology is immense, but in order for this technological revolution to benefit societies, not only of the West which gave rise to it but also the whole variety of non-Western societies, the artificial intelligentsia would need to take into account the specific cultural experiences of these other societies.

REFERENCES

1 Geertz, C. (1975). *The Interpretation of Cultures*. Basic Books, New York.
2. Popper, K. R. (1974). *Objective Knowledge*, Clarendon, Oxford.
3. Mitter, P. (1977). *Much Maligned Monsters*.
4. Curtin, P. D. (1964). *The Image of Africa*.
5. Vitruvius (1960). *De Architectura* (Translated by M. V. Morgan, *The Ten Books of Architecture*), Book VI.
6. Allport, G. W. (1954). *The Nature of Prejudice*.
7. Hochberg, J. (1972). 'The representation of things and people'. In *Art, Perception and Reality* (Eds E. H. Gombrich, J. Hochberg and M. Black).
8. Gombrich, E. H. (1959). *Art and Illusion*.
9. Oatley, K. (1978). *Perceptions and Representations*.
10. Bartlett, F. C. (1932). *Remembering*. Cambridge University Press.
11. Dilthey, W. (1976). *Selected Writings* (Ed. H. P. Rickman).
12. Wittgenstein, L. (1966). Lectures and Conversations on *Aesthetics, Psychology and Religious Belief*, edited by C. Barrett. See also discussion by R. Trigg (1973), *Reason and Commitment*.
13. Gellner, E. (1979). Concepts and society, in *Rationality*, Cambridge University Press.
14. Johnson-Laird, P. W. (1983). *Mental Models*.
15. Ayer, A. J. *Truth, Language and Logic*.
16. Feigenbaum, E. A., and McCorduck, P. (1984). *The Fifth Generation*.

Artificial Intelligence for Society
Edited by K. S. Gill
© 1986 John Wiley & Sons Ltd

10. THE ARCHIMEDES SYNDROME: CULTURAL PREMISES AND AI TECHNOLOGY

MASSIMO NEGROTTI and **DANILA BERTASIO** Sociology of
Knowledge, University of Genoa, Italy

Despite its numerous attempts or declared aims, AI has shown, up to now, only one reliable common denominator: the inferential one. The machine is still completely unable to reproduce knowledge processes properly defined.

One of the most important roots of AI tradition is just that of the knowledge concept. This paper discusses the problems that arise from a not-well-founded definition of knowledge and its relations with apparently similar terms such as 'reasoning', 'thinking' or 'symbolic processing'.

The need for deep researches on the 'cultural premises' of the AI community, in order to clarify both the social influences on its way of looking at man and the proper status that such a discipline could achieve, is also pointed out.

THOUGHT AND 'THINKING MACHINES'

The theoretical and practical aim of a thinking machine is proposable only if we define in operational and formal terms the concept of thought. This reduction is the most relevant constraint and result of the efforts that artificial intelligence (AI) has produced until today. Nevertheless, in the official literature of AI different terms appear which are often treated as synonymous, such as reasoning, intelligence, thinking, symbolic processing. The common denominator that it is possible to find in the real experimentations of AI is constituted prevalently by only one component of human thought, the inferential one.

After all, the thought that is reproducible through a program for a computer is in reality a logicomathematical 'calculation' and also more similar to reasoning. On the other hand, the etymology itself of the verbs 'thinking' and 'reasoning' is similar and connected to activities of measurement and of calculation rather than to the generation of concepts, ideas and intuitions that are the foundation of the critical activities of the intellect, *id est* of thought in its more proper sense. Theoretically, also, a machine could achieve formal reasoning even of a very high level, while it remains completely unable to reproduce knowledge processes properly defined.

This point of view is, on the other hand, intrinsically consistent with the nature of current hardware and of the programming languages—even the more evoluted, such as LISP or PROLOG. In each of these cases, like in that of the more conventional computer, what we have at our disposal is a complex architecture for the processing of information, not a process of knowledge. Unfortunately, the English term 'knowledge', which has no plural form, induces us to an improper use of the concept of knowledge, and therefore of thought. While expressions of the kind 'information technologies' are quite correct, others, such as 'knowledge engineering' or 'knowledge representation', seem to reinforce the unproposable image of a machine able to process knowledge, *id est* to simulate human thinking potentially in all its ways, formal and informal.[1,2]

In reality we should distinguish three levels very accurately: the brain (signals processing), the mind (intelligent processing of information according to established rules) and the intellect (knowledge activity in the Hegelian sense of *vernunft*, or 'reason'). For the moment, AI cannot claim to achieve successes at the third level, but only within limited ranges of the second, where, for instance, not only is it possible to process signals and information (or data) for normal uses but even to provide feedback for disadvantaged people, i.e. anywhere we have a formalizable problem to solve.

All the known programs of AI, in fact, are devices to enable solution of problems sufficiently formalizable; we know of no program able to generate problems. For this reason, the conflict, often very strong, between computer scientists and researchers of AI is partially unfounded. It could be different, and perhaps it will be, when AI would be able to build up a disciplinary foundation of its own, a paradigm of its own nature and of its own limits according to the knowledge available on the three levels we have proposed.

THE SEARCH OF THE CULTURAL ROOTS OF AI

When in 1980 we constituted, at the University of Genoa, the Research Group on Informatic and Cultural Processes, we were attracted by a thesis whose empirical references we have today in front of us. It consisted of the attempt to broaden the traditional definition of the sociology of knowledge, indicating as its task the study of social conditioning on thought, including the artificial one.

The devices coming from the computer field, in fact, rather than being considered technological or structural phenomena able to condition thought, should be perceived as objects which are themselves culturally conditioned, constituted by the thought and by the cultural premises of the societies they come from. Their influence or their impact on society is in fact to be conceived first of all as an interaction among cultural agents or models and not, as is typical for every other class of technological goods, only as an interaction between structure and culture.

There is an opposition, in this regard, between those people who consider the computer a pure tool and those that emphasize its 'thinking' power. This opposition is in a certain way similar to that stated by Searle when he speaks of a 'strong' AI and a 'weak' one.[3] As stated above, it is certainly to be excluded, for the moment, that it could be possible to attribute to any computer or program high mental or truly intellectual performance. Nevertheless, a computer is not to be considered like a lathe or a voice recorder.[4]

We could speak of a sort of 'Archimedes syndrome'. Just as the great scientist conceived his own inventions in terms of 'secondary and pleasant applications of geometry', in the same way several observers seem to consider the computer as a 'secondary and spectacular application of electronics'. Further, AI would be only a marginal application of computer science. Although they often incline to make spectacular forecasts about our ultracomputerized future, many of them do not understand, to paraphrase McLuhan,[5] that the true message of computer science and AI consists in its cultural nature and, in particular, in the prevalence that they give to the formal reasoning and to the supremacy that they induce of information with regard to knowledge, of the research of solutions rather than of relevant problems, of answers versus questions.

The same thing can be stated about the progressive substitution of classical references, such as the concepts, by terms that not even positivism has so much diffused, like that of 'data'. In its telematic version, the technology of information seems to promote, as pointed out in another paper,[6] a sort of 'data flow philosophy of life'. A life model within which the flow of data (i.e. of information), their possession and their exchange, should prevail not only over the flow of energies, including monetary ones, but with regard to the same speculative and intellectual activity, darkened by activities where 'to know' (in the sense of 'to know how' or to possess information) will be more important than the concept and the process of knowledge and 'to verify' by means of data will be more relevant than 'to perceive by intuition'. Similarly, to believe on the basis of informal persuasions, to doubt or to suspect would be culturally less approved than to simulate or to verify by means of a rational model.

On the other hand, the activities of computer programming require and encourage the exercise of our imagination and our ability in abstraction and the use of a symbolic approach.

As we can see, there are alternatives enough to be able to consider computerization, for instance in the educational processes, as a complex and delicate problem very far from a simple growth of the technical environment.

All this means that the whole question is together fascinating and crucial and it should be clear that it is a task of the sociology of knowledge to anticipate its dynamics. For example, it should be urgent to find, even through empirical surveys such as those that my group carry out on Americans, Europeans and Japaneses AI researchers,[7] the pragmatical roots of the sciences and technologies we are speaking about, indicating cultural disequilibria that should be controlled.

REFERENCES

1. Negrotti, M. (Ed.) (1984). *Intelligenza Artificiale and Scienze Sociali*. Papers by L. Gallino, A. Ardigo', M. Borillo, A. Sloman and V. Tagliasco. Franco Angeli Editore, Milano.
2. Bertasio, D. (1984). 'Expert systems and intelligenza quotidiana'. *Studi di Sociologia*, **4**.
3. Searle, J. R. (1982). 'Minds, brains, and programs'. In *Mind Design* (Ed. J. Haugeland), MIT Press, Cambridge, Mass.
4. Bertasio, D. (1984). *Educazione and Intelligenza Artificiale*, Li Causi Editore, Bologna.
5. McLuhan, M., and Fiore, Q. L. (1967). The Medium is the Message, Penguin, Harmondsworth.
6. Negrotti, M. (1984). *Cultural Dynamics in the Diffusion of Informatics*, *Futures*, **1**.
7. Negrotti, M. (1983–1985). 'How AI people think', Survey 8° IJCAI in Karlsruhe, Information Technologies Task Force, CEE, Brussels, 1983; Informatica e Documentazione, INFORAV, Roma, Vol. 2, 1984; 'Sociologia e ricerca sociale', Roma, 1985; 'How European artificial intelligence people think', Survey 6° ECAI in Pisa, ITTF, Brussels, 1985.

Artificial Intelligence for Society
Edited by K. S. Gill
© 1986 John Wiley & Sons Ltd

11. THE COMPUTER AS A CULTURAL ARTEFACT

BLAY WHITBY SEAKE Centre, Department of Computing and
Cybernetics, Brighton Polytechnic

My intention in preparing this paper was to provide an account of the computer, considered not primarily technologically but rather as a cultural artefact: i.e. as the product of a particular culture and as embodying certain cultural precepts. This seems to be an exercise which is rarely given serious attention and one which some of the speakers at this conference have suggested is clearly needed. Peter Large, for example, gave us a gentle warning about the way in which people have recently rediscovered the obvious truth that expert systems could contain and express the social and economic prejudices of their designers.[1] Several speakers at this conference have remarked on problems associated with taking the culture of users into consideration when designing certain types of computer-based systems (see, for example, Chapters 14 and 18). It seems reasonable that a designer wishing to take account of the cultural variations of users should have as clear a picture as possible of the cultural assumptions inherent in the technology itself.

There are other important justifications for the examination of computer technology as a cultural phenomenon. An extremely important justification is the way in which computer technology has wrongly acquired a 'value-neutral' image. This view is so prevalent that many people will have already reacted with surprise or disagreement to the implicit assumption that there are cultural assumptions inherent in the technology itself.

A common view is that technology is not to be in any way held responsible for the uses that are made of it. On this view modern computer technology is available for good or bad purposes alike. Whilst I do not want to assert the opposite extreme view—that technology is shaped entirely by social requirements—it seems reasonable to assert a middle view: namely that certain items of technology are more suitable for some social purposes than for others. The suitability of modern computer technology for use in particular applications depends not only upon the limitations of the technology itself but also on the cultural background of the technology. In particular, I wish to suggest that the applications for which it is suitable reflect the applications for which it was initially designed and, through this, the cultural assumptions of its designers.

It is worth pointing out at this stage that one's own cultural assumptions are generally invisible. The cultural assumptions with which one has grown up often seem to be just 'common sense'. They are only thrown into relief, that is they become visible, when someone from another culture challenges them in some way. If one has an interest in taking cultural factors into account in the design of programs or in constructing such devices as 'user models', it seems highly desirable to attempt to discover and examine the implicit cultural assumptions which one might be making and which might be involved with the whole computer paradigm. Consideration of the computer as a cultural artefact is an important exercise for those involved in AI and in the use of computers in education.

There is also a plethora of recent literature purporting to enumerate the social effects that technology, in particular what is usually called IT, is having and might have in the future on the cultures which are exposed to it. These range from the wildly optimistic, like that of Christopher Evans,[1] through simply optimistic and simply pessimistic to the gloomily pessimistic;[2] but for the most part they treat the 'information technology revolution' as analogous to an act of God (either a miracle or a disaster). It is something that is inevitable in that it will happen or is happening with its consequences good, bad and boring, rather as summer will give way to autumn in a few weeks. Of course, we all know that the progress of technology is not inevitable. Modern microelectronics is a human creation and any changes it produces in society are as a consequence of human choices. To remind people of this obvious truth and to assist in making those choices is one of the most important reasons for examining the computer as a cultural artefact. In order to remind people that they have choices with respect to certain technological developments my paper is deliberately provocative; I hope the reasons for this are now clear.

In order to consider the computer as a cultural artefact I shall not draw a distinction between consideration of the cultures which produced the modern computer and consideration of the sort of people who have become involved with them. These two factors are clearly related in practice and what is important in this paper is not to construct a sociological theory; it is rather to outline the sort of cultural assumptions which might be involved with the use of computer technology. In the time available it is only possible to briefly outline what I see as the major cultural factors associated with the recent development of computer technology. This, in turn, should provoke reflection as to the importance and desirability of working with those cultural assumptions. This task is not best served by conducting a survey of attitudes as Professor Negrotti has done (Chapter 10). In this paper I want to, first, present a very brief account of the cultural roots of the computer which have led to its cultural associations. Derived from that I want to suggest the existence of a 'computer culture'. I will also briefly attempt to identify a computer 'counter-culture' as a subversive movement within this culture and to show that these issues have genuine practical implications.

THE CULTURAL ROOTS OF THE COMPUTER

Apart from Babbage and Lovelace who, it seems, made the familiar British mistake of being the best part of a century ahead of their time, it is generally agreed that the modern digital computer has its roots in the Second World War. The Bletchley Park Colossus machines, which were designed and built during the Second World War for the purposes of mechanizing the deciphering of coded transmissions, provided most of the impetus for the ideas that grew into the modern digital computer (see Chapter 4). The first Manchester machine involved two of the key figures from the secret wartime work at Bletchley Park: Professor Newman and Alan Turing. At the same time work in the United States on the ENIAC machine, which also stimulated many developments in computing, was carried out for the US Army Ordnance Department.

During the Second World War there was also considerable improvement of, and need to improve further, analogue devices which predicted the courses of aircraft and the trajectories of guns and missiles. The increasing use of radar required new and improved communication and control networks. These needs were all intimately involved in the creation of the computer as we know it. The original inception of the computer was to meet military needs and it is fair to say that all subsequent developments have been led by military demands. In every major computer manufacturing country apart from Japan the overwhelming proportion of research funding comes from the military, either directly or indirectly.

The connections between the military and the computing industry are frequently remarked. The relevance of these connections to the cultural roots of the computer lie in the continuing influence that military requirements have had on the culture associated with computers. The requirements of military operations can be clearly and briefly stated as more or less those listed below (Table 1). The nature of military operations, throughout history, has required fast and accurate information handling. The penalties for any departure from the optimum level of information handling are frequently swift and dramatic in a military context. The charge of the Light Brigade is a good historical example of this. The present developments of information technology have gone a good distance towards meeting these military requirements. An interesting addition to military requirements is provided by the fact that the discovery of nuclear fission has in turn produced a more general and crucial need for fast, accurate and reliable command and control systems.

Table 1   Features of military operations

1. Clear objectives must be stated.
2. Speed is essential.
3. Accuracy in exchanging and processing information is essential.
4. As much information as possible must be gathered.
5. As little information as possible must be given to the enemy.
6. Mechanisms involved in the sending and receiving of information must be totally reliable.
7. Total obedience is required in all operations.
8. Command and control must be structured and rigid.

The computer has not remained an exclusively military development. Although the connections with military needs are historically important and as relevant as ever when we look at present-day sources of research funding, particularly for work in AI, the computer has been effectively civilianized. During the period since the Second World War, which I have claimed represents the cultural roots of the modern computer, many important developments have been made in response to non-military needs. What is important about these needs and the developments made in response to them is that they are not different in kind from the military needs listed in Table 1.

The needs of large commercial organizations and of large organizations of the state are essentially similar to those of the military, particularly in respect of the need for fast, accurate and reliable command and control systems and efficient information-gathering networks. These organizations were therefore quick to adopt and use the computer for civilian purposes as it became more reliable and less esoteric through the 1950s and 1960s.

The civilianized computer, however, remained the property of large organizations and, significantly, the property of the advanced industrialized nations. With the possible exception of Russia, all these nations shared a cultural adaptation of at least a century to a society based around the needs of high-output mass-production techniques. These mass-production techniques provided both the capital for further computer development and the cultural background for civilian development. The role of the computer in 'deskilling' therefore does not need to be seen as a suprising technological consequence; rather it is an expression of social needs which were already in existence in the cultures which produced the civilianized computer.

THE COMPUTER CULTURE

From these roots a particular set of cultural assumptions appear to have been developed. Essentially the computer has become very effective at meeting the requirements in Table 1. These requirements probably look rather familiar in that they are more usually presented as the sort of fields in which computer technology can be successfully applied. To some extent these successes may be regarded as limitations of the technology. An alternative explanation may be provided by the route taken here. This list is not derived from any observation of what computers actually do in 1985, but rather from an account of what the computer-building culture wanted to do. By now computers can actually do most of these things quite well. The problems discussed in this volume have frequently been concerned with the difficulty of making computers do anything else. As an illustration of this consider Partridge's list of the successes of AI (Table 2) (Chapter 4).

Table 2   Achievements of AI

1. Chess-playing programs	Success
2. Theorem proving	Success
3. Natural language understanding	No success as yet
4. Expert systems	Some tentative success

It is not difficult to see how these successes and failures can be explained in terms of the cultural antecedents of computers. An important observation is that chess is one of the oldest war games, and since the nature of military operations have not changed since the time of Julius Ceasar, it should come as no surprise to us that a piece of technology demanded and paid for by military interests can be taught to play a war game. Eloquent statements of the pacifist position have been made. I want to encourage reflection upon the possibility that the fact that a piece of technology, built in response to military demands, can easily be programmed to play a war game whereas understanding a natural language such as English seems much more difficult is perhaps not suprising from a cultural point of view. The pacifist ideals may run contrary to the ideals of the culture which originally built computing machinery.

Many of the problems which appear to be technical problems might well be better considered as problems which stem from trying to use a piece of technology which was designed with military requirements in mind. The original computers were produced by cultures undergoing a worldwide war and have subsequently been developed by the main participants in a continuing worldwide cold war. That is not to say that all the problems which AI researchers face can be explained by reference to the influence of the military. It seems that there is a good case for at least examining this possibility further, particularly in the case of AI, which appears to be experiencing a slowing down in research results at the same time as it is experiencing an increase in interest in military applications. Similarly, the familiar complaints of educationalists—that too many computer games involve shooting at things—can be easily explained from this examination of the computer culture. Computers are technically suited to such games because the original machines were designed with the calculation of trajectories in mind. The sort of person attracted to work with computers and the developments which seem attractive may be further influenced by military demands. In the case of the development of, and skill in the use of, shooting games the interests of the military and all computer users are convergent.

What then is the 'computer culture' like? It is really most like a military culture. It has adopted a set of working practices which conform well to the list of military requirements in Table 1. One might even claim that the desirability of total obedience has been adopted by the computer culture. Weizenbaum suggests that one of the reasons someone might become a compulsive programmer is that a programmer has total and complete control over his/her program.[3] The expression that Weizenbaum uses is 'omnipotence'. Software engineers regard disobedient programs as very dangerous things (aside perhaps from the extreme fringe of AI work). Is it too much to suggest an analogy between a faulty procedure and a deserter?

The computer culture has had its greatest successes in areas where clear goals and the means of obtaining them can be defined. It is also very capable of handling, storing and transferring vast quantities of information quickly and accurately. It is true that these abilities of computers are valuable in other application areas. Databases containing patient's records and medical files, for example, require the highest level of accuracy and reliability. The main argument here is not challenged by this. The cultures which developed computer technology had a need for these technical abilities in a military context and when considering the computer as a cultural artefact it is the military demands which seem to be the motivating force and the ability to handle large numbers of patient's records which seem to be seen as a 'spin-off'.

There is not really time here to explore fully the consequences of the fact that computers have been developed by cultures which have a history of mass-production techniques and large-scale industrialization. A similar argument to that about the influence of the military culture can be made about the influence of figures such as Taylor and Ford. In brief we can say that it is not surprising that computer technology is often accused of 'deskilling' workers. This is just the sort of role that the culture which originated and developed computer technology had in mind in its original development of the technology.

There are two other important points to be made on the computer culture. First, during the Second World War, British government propaganda put a lot of work into building up the image of the specialists who worked in code breaking and scientific intelligence. These were the original 'backroom boys', whose intellectual prowess could be used to achieve socially useful objectives, in this case those of winning the War. The military virtues of the previous generation were still mainly those of courage in battle, but during this period the propagandists made a virtue out of technical expertise. The public was told that just as important as the brave soldier were the technologists breaking new ground in developing radar or the intellectuals cracking the enigma code.

This public image is still having repercussions today for the computer culture. A particularly obvious example is the case of the computer criminal. The computer criminal, according to most accounts, has a very positive public image. This is, on the superficial level, quite remarkable. If someone robbed a bank by violence or by embezzlement, they would probably be held in fairly low esteem by the general public; yet if they can use a computer to do it they are perhaps even admired. Most computer crime (various authorities have put it between 60 and 80 per cent) is committed by trusted employees of the organization that is robbed. A fair comparison of the typical computer criminal therefore would be with a shop assistant who cheats the till or the clerk who fiddles the books. Given this comparison it is difficult to see why there should be such admiration by the public. It is highly likely that this is a hangover from the wartime image of the bright young man who could crack the code.

This high public valuation is not contradicted by claims, such as those of Turkle, that 'hackers', today's inheritors of the position of the wartime 'backroom boys', have a low self-evaluation.[4] The high regard which the public have for computer criminals is not necessarily rational, since it was engendered by the efforts of wartime propagandists. Also Turkle is more concerned with the self-image of the sort of people who become hackers. The image which the public have of such people is quite possibly more positive than that which they have of themselves. Whatever the psychological mechanisms which motivate people to become part of the computer culture today, in the eyes of the public they are part of an elite. I have examined the problems of elitism in this area elsewhere,[5] so I will not expand on the subject here.

The second observation on the culture of the computer concerns those insights into the cultural assumptions of those associated with computers by examination of the sort of language which they use. Computer jargon is such a rich field that observations and discussions of particular terms could continue indefinitely; for present purposes it is sufficient to consider a couple of relevant examples. First, the use of the expression 'coding' is obviously consonant with what I have claimed about the origins of modern computers in code-breaking machines and the origins of 'hackers' in the 'backroom boys' during the Second World War. Even writing in a high level language is sometimes called 'coding' by programmers—perhaps trying to claim some of the positive valuation resulting from the efforts of the wartime propagandists.

A more interesting case is the use of the word 'powerful'. This is clearly not a literal use of the word. When we say that one computer is more powerful than another we do not mean that it generates power in the strict physical sense. Nor does it mean that a computer exerts more power over its environment in the sense in which we might apply the description to a weapon or tool. In fact, 'powerful' when applied to computers means something very different, something closer to 'capacity'. Not strictly capacity, perhaps, since there are considerations of adroitness in handling data as well; however, 'speed' is usually cited as a separate attribute to 'power' in this context. 'Powerful' is a choice of term which suggests a great deal about those who choose it. It is a word which is frequently applied to weapons. Not only weapons, but tools and weapon-support technology are often called powerful—telescopes and searchlights are obvious examples—presumably because of the power they give their user.

If a term such as capacity had been chosen instead this would suggest something much more passive, more feminine, perhaps. When we ask why certain people, and women in particular, find work with computers less attractive, it is worth reflecting on the cultural assumptions that are conveyed by the use of a word like powerful in this context.

The existence of these cultural assumptions has particularly important consequences for research in AI. Several researchers have remarked on problems which stem initially from trying to work at a level of accuracy and reliability which is inappropriate for the domain. Work in AI is frequently involved in attempting to model human decision making which is imprecise and error prone. Working within the culture of the computer may lead one automatically to the assumption that computer-based decision making must be precise and error free. That is not to claim that all software standards should be relaxed for research to be fruitful. It is worth making the point, however, that some of the standards which are expected from programs may be derived from a computer culture which has unquestioningly accepted standards appropriate to military operations (Table 1). Some imprecise and error-prone systems, often systems involving human beings, can achieve good results in some situations. If AI researchers want to reproduce these results they may find it useful to relax some requirements which may be culturally produced rather than appropriate for the task.

THE COMPUTER COUNTER-CULTURE

Computers have also been used in a way that is, with reference to the cultural assumptions described so far, truly subversive.

The most general subversive use of the computer is in education. This is not a use of the computer which follows directly from its cultural roots; nor is it in any way compatible with the assumptions of the computer culture so far described. The computer was designed and built to serve as a control, planning and memory aid for fairly sophisticated and highly trained minds. The fact which is now apparent is that it can be used to help develop the minds of those who are neither sophisticated nor trained is something that has emerged from the use of computers in a way contrary to the demands of the mainstream computer culture. The demands of the mainstream computer are still related to the needs of the military and of large-scale business. Alternative applications of the computer have been demonstrated by a fairly small group of dedicated people.

That these applications are subversive, with respect to the computer culture, can perhaps be seen by briefly examining some of the cultural currents that surround the much vaunted use of microcomputers in British schools at present. Parents, government, business and even military interests are persuaded to accept or encourage the 'computer in every school policy' because it will make our young 'computer literate'. Computer literacy is not the basic life-skill that the use of the word 'literacy' implies here. Computer literacy can mean various things to various people. To ambitious parents it can mean that just because their children have spent some time playing with a microcomputer, they now have the ability to enter the elite—the highly paid computer professionals who still carry the positive valuation from the days of wartime propaganda. To government, industry and the military, the increase of computer literacy means that there is a large pool of available labour which has at least a grounding in the skills which they might perhaps at some point in the future require.

What educationalists and we at the SEAKE Centre are trying to do is quite different. We are trying to use the computer as a tool to educate—to promote the intellectual development of the individual for its own sake. In this aim we are most grateful to those who have seen fit to provide the equipment.

In considering the counter-culture in computing, it is worth mentioning that the very existence of microcomputers is a consequence of subversion. The large-scale manufacturers of computers would have preferred to exploit technological advances by giving the same customers more computing power (to employ that term for once) at the same price. Greater centralization is after all a part of the mainstream computer culture. It is the small-scale manufacturers who have forced IBM's hand in producing the PC, and already the economics of giving computing power to the masses by selling home computers looks very shaky. The usual case against providing computer power for home use has been that it is not profitable. However, this is no more than an economic expression of the fact that the more powerful institutions in society can pay more for a useful product than those without. There is no doubt that the vast changes in the cost of computer technology will make the economics of selling home computers vastly more attractive in future. For the present purposes it is only necessary to observe that it is probably not a coincidence in cultural terms that most home computers have been produced by small organizations and by individuals dropping out from the mainstream computer world in order to produce micros for home use, either out of social conscience or the desire to make a quick personal profit. There is no obvious economic reason why this should be so. The large computer manufacturers had the advantages of experience and of economies of scale. It may well be that the large manufacturers were unwilling to take the risk of entering this market. These risks are obviously very real as the recent downturn in the market had demonstrated. Within the British financial system it is easier for large companies to enter risky markets than it is for small companies. The introduction of home computers has applied a mild shock to the mainstream computer culture and raised some interesting questions about the distribution of information and power within society, particularly since the elitist mystery of the computer is undervalued if everybody has one at home. It is, however, a shock from which it will, in all probability, rapidly recover.

REFERENCES

1. Evans, C. (1979). *The Mighty Micro*, Victor Gollanz, London.
2. Randell, B. (Ed.) (1982). *The Origins of Digital Computers*, Springer-Verlag, New York; especially the papers by Randell, B., Juley, J., Rachman, J. A., and von Newmann, J.
3. Weizenbaum, J. (1984). *Computer Power and Human Reason*, Chap. 4, Pelican, Harmondsworth.
4. Turkle, S. (1984). *The Second Self, Computers and the Human Spirit*, Chap. 6, Granada, London.
5. Whitby, B. (1984). 'AI: some immediate dangers'. In *AI: Human Effects* (Eds, M. Yazdani and A. Narayanan), Ellis Horwood, Chichester.

Artificial Intelligence for Society
Edited by K. S. Gill
© 1986 John Wiley & Sons Ltd

12. ART AND DESIGN: AI AND ITS CONSEQUENCES

GRAHAM J. HOWARD Department of Art, Coventry Polytechnic

INTRODUCTION

Information technology has again brought the visual image to the fore in the context of our, essentially text-based, society. The implementation of AI in information technology has raised the problem of how we, ourselves, understand and use images. Artists and designers have a wealth of experience of images and their function accumulated over hundreds of years. A dialogue, concerned with the nature of image understanding, image use and the social and political implications of images, should be occurring between the AI community and those artists and designers involved with the use of information technology. This dialogue would benefit both the AI community and the artist and designer. Images are powerful, subtle and dangerous weapons; they alter how we see the world. Their revitalization by information technology, further empowered by AI, is an exciting yet daunting prospect. Discussion of the social and political consequences of AI in image understanding and production is long overdue; at present the silence is deafening.

INFORMATION TECHNOLOGY AND ART AND DESIGN

Artists and designers have become familiar over recent years with information technology where it involves image production. Video, especially, with its broadcasting and industrial applications has increasingly required

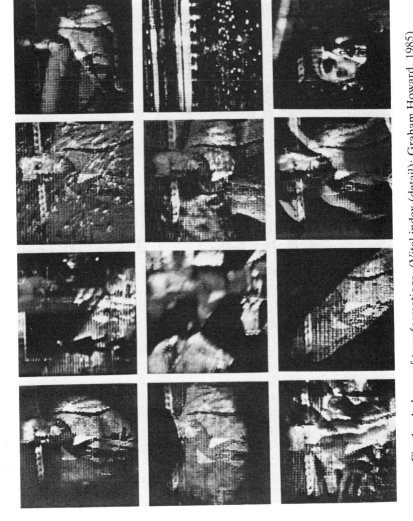

Fig. 1   A dance of transformations. (Vital index (detail): Graham Howard, 1985)

their skills. Computer-driven technology is fast providing a new set of manipulative tools for the artist and designer. The usage of these new tools is being taught to art and design students in art schools, colleges and polytechnics. These new tools, computer-aided art and design (CAAD), include CAD, page formatting and paint-box type systems. They have increased the range of possibilities in image production both by the nature of their ability to configure complex images, as in Fig. 1, in ways that were previously very time consuming and difficult and also by their speed of operation. This has already led to a minor explosion in the number of images produced by artists and designers in, for instance, television current affairs programmes. Indeed, we now see the products of these systems every day in advertising, magazine layouts, television news presentations, etc.

However, although we see the products of this technology we are not normally aware that this is indeed what we are seeing. Most of the final artwork is not visually distinguishable from more traditionally produced artwork. This might be said to be one of the criteria by which such production methods may be judged—that the final product should not conspicuously show the nature of its own production whether it be cut-and-paste technology or whether it is sophisticated computerized versions of this interfaced with video digital frame-storing techniques. This hidden aspect of the technology is important for the way in which we consume the images produced: the images must convince. Further to these techniques there is the rapidly growing area of computer-generated three-dimensional modelling and animation. At present this is only rarely capable of being produced so that its ontogenesis is hidden, and for the most part the unnatural or mechanistic quality of its visual product is celebrated. With the use of fractals and other devices this will soon change. We are at a point of crisis with respect to image production and consumption: early printed books looked like manuscripts; initially print culture imitated scribal culture and now incipient electronic culture imitates print culture. Information technology and CAAD are being developed and improved all the time and because of the nature of the tasks they are being employed to do, financial investment for this research and development is forthcoming. The intersection of CAAD with IT is a major growth area which will have enormous impact on our society. The growth of expert graphic systems and the intersection with AI will accelerate this impact.

Images are complex and highly context related; the theoretical difficulties surrounding their analysis are often very similar to those surrounding the analysis of natural language. The labyrinthine, encyclopaedic quality of natural language has caused problems to the AI researcher for years; the image is even more recalcitrant and far less work has been done.

DIDACTIC, MNEMONIC AND PROPAGANDIC IMAGES

Before considering further the problems surrounding the intersection of art and design with AI it is important to stop and consider how the production and consumption of art and design has functioned in the past and how this historical context may allow us some useful insights into the analysis of images. The following are historical attitudes based upon a view of art and design practice that is Western European centred; how this historical view is extensible to other cultures is beyond my scope and significantly so. These are culturally and historically determined remarks by necessity.

Didactic Images

Mediaeval mural painting forms a useful starting point in understanding both the internal structuring of images and the social and political significance of their use. These are images produced at a time when orality was the dominant form of communication but was starting to be changed by the impact of text-based communication. The full force of this change was not to be felt until text's most potent implementation in print technology totally restructured our conception of the world.[1] They are essentially didactic images, used both for the propagation of information and for the memorization of information. They are not without contemporary parallels. Mediaeval church wall paintings, as in Fig. 2, were designed with the intention of telling the illiterate and the semiliterate the essentials of what was in the Bible and allowing for the elaboration of these by the priest whose pulpit was usually situated either by the side of the mural or in direct opposition to it. This positioning allowed the priest the possibility of using the images when preaching as a device to focus the attention of his audience on the topic with which he was particularly concerned. There is a clear situational similarity both to the configuration of television news production with its use of the newscaster juxtaposed against the electronically derived image depicting some element of the news topic, as in Fig. 3, and to the configuration of the classroom teacher using text and the drawn image on the board or on the overhead projector. In this manner the priest fulfilled both these functions, for the preacher was also the disseminator of local news and official government information. The mural images, because they were static and semipermanent, had to be very rich in informational terms in order that a wide variety of sermons encompassing many different moral lessons could be preached using many different texts and that these texts could be indexed by the priest to the image so that the audience on seeing again the image would remember the texts. The image in Fig. 2 is of a Last Judgement, showing Heaven and Hell, accompanied by, probably, a Seven Virtues and a Seven Deadly Sins. These images were designed to allow belief structures to be fixed in people's minds; the belief crept in via the image and consequently became intimately associated with it. This method of using images presented the world in a particular way to its audience and therefore helped to structure the audience's view of the world. The element of control is important here; images were used to limit the way in which people might view the

Fig. 2 *Last Judgement*, St. Nicholas Church, Oddington. (Photo: Graham Howard)

Fig. 3   BBC News. (Photo: Graham Howard)

world and actions within the world. If you did not follow the moral imperatives of the ruling orthodoxy and contravened their regulations, then the inevitability of the soul's burning in hell was graphically clear. Those who obeyed and upheld the beliefs of the dominant were rewarded with the imaginative visual delights of heavenly peace. This technique of the simple juxtaposition of opposites is still with us and is used daily on our television screens by advertisers. Didactic images often become propagandic images.

Mnemonic Images

Other images functioned formally in a similar way but were specifically designed to allow a precise textual memory to be associated with an image. These images were invented and manufactured to memorize long passages from the Bible, sermons and passages from classical texts. In other words, they were devised to allow those who could read a way by which they could extend this ability in the face of the scarcity and expense of books in a scribal culture.[2] The form of these devices varied greatly: some were painted images specifically constructed to serve as memory devices, others were objects of the imagination fixed in diagrammatic form, others used the interior of buildings, especially churches, as a means to locate memories. These allowed the person recalling the memoery to walk around the building reconstructing the memories in correct sequence as they perceived each part of the building with which they had associated a particular piece of text.[3] In an essentially oral culture Thomas Aquinas suggested, 'Simple and spiritual intentions slip easily from the memory unless joined to corporeal similitudes'.[4] He further advised that these similitudes should be unusual. This recommendation led to images either having within them some bizarre or grotesque element or an internal incongruity between juxtaposed parts. The type of images used came to change over the years from the scholastic emotional, dramatic and corporeal similitudes, through to the Lullist devices which were characterized by their use of letter notation and movement, and eventually to the techniques of the Ramist memory system which were based on dialectical order and heavily influenced by the rapidly advancing print technology. These Ramist devices were influential in the production of early learning devices for children, forming the basis for the first printed ABC learning books and fundamentally affecting our approaches to educational methodology. All of these images, whether derived from hermeticist, Kabbalistic or Ramist viewpoints, allowed knowledge and belief structures to become embedded in people's minds. In many ways the Ramist method, which was developed under the influence of the implementation of print technology, is still utilized by teachers now. It is heavily text based, associated with analysis and rationality, and forms the hidden infrastructure of most present uses of videotape and videodisc technology.

IMAGES, KNOWLEDGE, BELIEF AND DOMINANT DISCOURSES

Historically, images have been used for mnemonic, didactic and propagandic purposes. Images have been used to remember, teach and propagate knowledge and belief structures and their associated embeddings in contemporary technologies. Further to this, they have been used to generate, confirm and change knowledge and belief. Although images have previously been essentially configurational, even the introduction of technology allowing sequential formatting of images and their conjunction with other means of communication has not significantly altered the way in which they have been made to function.[5] Film, television and computer technology have used them in fundamentally similar ways and for fundamentally similar purposes.

The way that images look is dictated by their subject matter and the subject matter of images tends to be dictated by the dominant discourses of the time.

The intimate relation of the image with the dominant discourse of the time is crucial to its function. The image becomes an icon of the belief and knowledge structures assumed by the dominant discourse. The icon is understood as a focus of this knowledge and belief: in some difficult and complex way the knowledge and belief structures inhere within the image as the icon. This is a significantly different definition of the icon as understood within current semiotic theory where an icon is seen as an image that physically resembles its referent.[6] The icon may be a crucifix, a map, a car, the image of a film star, a family snapshot or many other things.[7] The icon is an image that carries with it its own epistemic and doxastic baggage and only when some catachresis occurs is it separated from this baggage: this separation is always significant and related to a historic shift in knowledge and belief structures. The catachrestic nature of many signs, like the Christian cross or the Nazi swastika, is symptomatic of the way in which visual images are altered both by context and by text. The reversal of meaning can occur either by the influence of context or text but is most clearly seen in the relation of the visual image with text. In some ways the change in text encapsulates a change that may take hundreds of years with respect to a change in context. In two consecutive sentences the meaning of the cross can be changed from an image of the death of a disgraceful criminal to the image of the son of God; for this image to change in the context of society took hundreds of years. Of course for the change to occur in context requires the prevailing dominant knowledge and belief structures to be changed, and this is always a considerable task—a task in which text in a literate society can have a powerful and effective role. It is clear that the change of meaning of images can occur much more quickly in a literate society than in an oral culture and, further, images may change their meaning even more quickly in an age of electronic communication. That this catachresis can be effected in a much faster time implies that the knowledge and belief structures of our society are available for change at a far greater speed than previously. The implication here is that the electronic culture will be littered with epistemic breaks and doxastic fractures unless there is a reversal in the social and political pressures which encourage change and the development of the new. In times of trouble and crisis the dominant discourses and their associated knowledge and belief structures are threatened. The nature of the crisis will determine the scope and direction of the threat and also the particular knowledge and belief structures subject to change. As such changes occur so the icons associated with these structures become subject to change also, and indeed often become a focus of these changes. The consequence for the icon may either be catachresis, where some radical shift in meaning takes place, or iconoclasm, where the icon is broken coincidentally with the epistemic break and/or doxastic fracture. When the latter occurs it is normally followed by a period during which other images are gradually developed and are asserted as icons of the new predominating ideologies.[8,9]

The critical discourses of subdominant groups tend not to produce images that function as icons, but these discourses attempt to either force a catachresis or to be involved with a visual iconoclasm. Historically this is the situation where cartoons and caricatures have features. Such images form the visual cutting edge of critical discourses and aim at wounding the dominant discourses from within. They either transform the icons of the dominant group into the opposite of their original meaning or they deconstruct and destroy the icons. The former is catachrestic change and is characterized by the carnival and the world turned upside down, and the latter is visual iconoclasm and is characterized by the visual cancelling of the dominant icons and is often reinforced by the physical destruction of these icons.

The discourses of other subdominant groups may produce icons, but in a situation where these icons are embedded within a discourse that is outside the domain of the dominant discourse and so although opposed to the dominant discourse does not come into direct conflict with it and its icons. This situation allows for the survivial of both sets of discourses and their icons without obvious damage, even though the discourses may be mutually contradictory. How this occurs is subject to the nature of the groups involved and to the significance of the contradictions. If the subdominant group is not seen to be threatening to the status quo, its critical discourses can either be regarded as harmless or rendered harmless through condescension. In any society the dominant discourses endeavour to reproduce themselves and propagate their attitudes through the production of icons, and in doing so attempt to and often succeed in silencing others. This silencing may take many forms ranging from heavy-handed censorship to some very subtle and insidious methods where the meaning of images important to the subdominant group is denied. This denial is often effected by a form of cultural hijacking. Such cultural hijacking can lead to full-scale cultural imperialism.

The dominant discourses of a society are naturally reproduced in the dominant technologies of that society and are intimately linked with it. In our society print and information technology are the powerful means of implementation of current discourses. Although text and its printed form underlie much contemporary use of computer and video technology and therefore may still be seen as the most influential shaper of our world, it is clear that the conjunction of the visual, oral and literal in sequential form in the new electronic technology furnishes us with a new, complex and potent weapon. Advertising, whether commercial or political, has rapidly begun to exploit this potential on television and we are just becoming aware of the significance of the new computer technology, especially when it is configured by AI. Much of the power of these systems will derive from their use of images in both configurational and sequential forms and this potency is gained from our historical understanding of images. Any attempts to produce systems that do not take into account the nature of image understanding or assume that it is simple and well understood are doomed from the outset. Because we are still reliant upon a text-based view of the world there is a persistent simple-mindedness apparent in the AI community with respect to image understanding; this is particularly clear in the use of images in AI and in the tendency to ignore the richness and power of images in our world. The didactic, mnemonic and propagandic images of the past are the foundations on which our present understanding and use must be based. All this is to suggest that the production and consumption of images is a good deal more complicated and significant than might appear at first glance.

AI AND ART AND DESIGN

Expert Graphic Systems

Expert graphic systems derived from current technology in use in the area of art and design will soon be producing images that will appear in print and video form. Such systems will enhance the ability of the artist/designer to manipulate visual elements both within a configurational and a sequential format, and further to this, allowing the artist/designer access to methods of visualization employed in the past by both herself/himself and other artists/designers. Many of the traditional manipulative skills could be lodged in a system in this way. A system like this would function solely as an expert system helping the artist/designer to carry out various manipulative tasks, but its effect on the nature of the product must not be underestimated, especially within the social and political domain. For instance, the ability to alter and materially affect news images in a sophisticated fashion that would disallow detection after the event is clearly an ability that would require treating with some caution. Such a machine could facilitate such alteration of not just static images but also moving video images, and make the activities of Goebbels and Beria look like child's play. If this technology was used in a blatant fashion then critical groups would soon become vocal, but the technology will enable very subtle things to be done with visual images and in this lies danger. Imagine the situation where the image of a person in political power was always subtly enhanced to look benevolent and attractive and the image of another was always subtly twisted or exaggerated to look slightly grotesque. Clearly this would be a subtle and devastating weapon in the arsenal of the manipulator. The basic ability to do this is available now but requires a great deal of time and skill; the expert system would make this quickly and easily available and therefore a viable possibility. In a parallel circumstance, books were in existence long before printing enabled them to make their full impact. As has been said above, dominant discourses endeavour to reproduce themselves and dominant groups tend to control the dominant technology. Expert systems of this kind could easily take their place in this domination. Eventually these systems may be utilized by subdominant groups to produce an effective critical discourse. These attitudes and concerns reflect the attitudes and concerns of those disturbed by the political and social consequences of the implementation of print technology before print had become a dominant means of communication.[10,11] At that time the concerns were with the religious and political power of the Church and the monarchical state;

now the concerns are with the commercial, financial and political power of multinational companies, international banks and national governments and their agencies. The subdominant groups now tend to be world wide but their vulnerability is just the same. A keen awareness of the nature of the funding that AI research attracts is of great importance, first to the AI community but crucially to those in the subdominant groups in society.

Intelligent Image Producer

The construction of an intelligent image producer poses a problem of a different order. Any intelligent image producer would have to be an intelligent image consumer as well: i.e. it would have to be capable of understanding images in order to intelligently produce images. The level of understanding of image that is required is over and above the levels of Marr's schema for computer vision.[12] Marr outlined a framework for visual information processing that included three levels: (1) the primal sketch; (2) the two-and-a-half-dimensional sketch; (3) the three-dimensional sketch. For present purposes a fourth level would be required: image understanding. Image understanding would involve the location of the image in the context of knowledge and belief structures; it would require the specific elaboration of its context and at least some of its potential contexts. How such an image understanding device might process and locate the images that are made available to it is clearly very important. Given the significance of the conjunction of image and text in our society one of the most obvious ways of enabling image understanding to occur is to elaborate a process whereby a pictorial database is indexed to and cross-indexed with a semantic database.[13] It is crucially important how such an indexing occurs within a particular machine. The possible number of interpretations of the conjunction of an image with a piece of text and vice versa is obviously large unless the domain of discourse is restricted. The complexities and difficulties of indexing may be shown by the example of Fig. 4; this photograph was taken during the 1984–85 miners' strike and attracted a good deal of attention. The way in which different social groups indexed this photograph could be set out roughly as follows:

1. This is an image showing typical police activity during the strike and it is good to see such images appearing.
2. This is a very disturbing image which shows the police in a new and frightening light.
3. This is clearly a fake designed to incite anti-police feeling.

The social group that has control over the indexing of such an image is in a powerful position with respect to how this image might be used. At present such indexing occurs in a loose and comparatively unrestricted fashion, but if it is built into an intelligent image producer dangers soon arise. Further, this photograph is now public and well known and is,

Fig. 4   Orgreave, 18 June 1984, miners' strike 1984–85. (Photo: John Harris, IFL)

therefore, restricted in a relatively clear manner. Its context and the surrounding discourses are open for investigation: it is a simple, if controversial, example of a very complex problem.

As the number of distinct interpretations of an image increases and the complexity of each interpretation increases so the complexity of the indexing task and the amount of computing power required increases. In order to restrict the amount of computing power that would be required to deal with vast arrays of interpretations and associated indexings inherent within a wide-ranging system, the domains of discourse would have to be radically restricted. How this restriction delimits the domains of discourse depends on the task in hand or on the dictates of the client for whom the machine is constructed. If the machine is to produce weather maps, it is reasonable to leave out those domains of discourse concerned with philosophical speculation.

Given that the most likely task for the artist/designer, to whom such machinery might initially be made available, will be the reproduction and propagation of the icons of the dominant discourse, the most probable way in which the domains of discourse will be restricted will be to include only those discourses that are encompassed by the dominant discourse and do not run contrary to it. Put another way, images will only be 'seen' by the machine by the way in which they are mapped onto the dominant discourse, and any contradictory interpretations will be eliminated from the start as mere fictions and not 'objective' interpretations of the image. This would result in the cleansing of images generated by the machine. In some social and political climates such a machine could be used for the processing of news information before broadcasting and, conjoined with rigid censorship, would form a formidable opponent for any dissenting group. Interestingly, this again parallels the situation when print technology was first introduced. The conjunction of image and text in printed and, therefore, distributable form was used by the dominant groups to restrict and launder information. The cleverest, and in consequence the longest lasting, of these groups used the print technology to colour and adjust the focus of information in a subtle and sophisticated manner.

Censorship and image cleansing are symptoms of, and result in, complete political and epistemic conservatism; this is clearly unacceptable to subdominant groups, politically, and to those involved in the growth of knowledge, whether they are educationalists, artists or scientists. Fortunately this outcome of attempts to produce and utilize an intelligent image producer is not a necessary outcome, but it remains a possibility if AI researchers and the general public do not recognize the potential power of images and their elaboration in AI-driven electronic technology. However, much of the socially manipulative power is derived from the potentially exclusive use of the equipment by those in present positions of power, and this exclusive use is in turn derived from the cost of development and production of this machinery. Printing presses with movable type were also very expensive and initially the property of those who were already successful and rich. Gradually costs came down and presses fell into the hands of those who were critical of the dominant discourses. These new owners of the presses formed a new social group; enabled by print technology to communicate new and revolutionary ideas they became influential in bringing about radical social, political, scientific and technological change. Although present computer technology upon which an intelligent image producer might be based is very expensive and unlikely to be available to subdominant groups, this is unlikely to be true in the future.

The above discussion has tended to concentrate on the dangers of the implementation of expert graphic systems and intelligent image producers. We must not ignore the tremendous possibilities of such technologies if they are used to increase our knowledge of the world and ourselves through a reassessment of how images may be understood and used. For this to occur the machine would have to suggest new and significant interpretations of images, ranging over wide domains of discourse, and be capable of understanding and producing interesting visual and verbal fictions. A machine with this capacity would have worried Plato, just as the artist and the poet did.

CONCLUSION

The impact of AI on art and design is bound to be large and significant. Expert graphic systems will be with us soon; intelligent image producers are probably some way off. Artists and designers will have new tools with new potentialities that will lead a reassessment of the robe of the artist and designer in society. The social and political consequences of the implementation of this technology are going to be with us sooner than many might think, and it is vital we address ourselves to these complex and difficult issues of the image in the electronic age as soon as possible. The power of images embedded within significant discourses will not disappear because we are not attending to it, and dominant discourses have ways of embedding themselves.

REFERENCES

1. Ong, W. J. (1982). *Orality and Literacy*, Methuen, London.
2. Eisenstein, E. L. (1979). *The Printing Press as an Agent of Change*, Cambridge University Press, Cambridge.
3. Yates, F. A. (1966). *The Art of Memory*, Routledge and Kegan Paul, London.
4. Howard, G. J. (1971). 'Revelation and art'. *Art-Language*, **1**(4), 6–16.
5. Nadin, M. (1984). 'On the meaning of the visual'. *Semiotica*, **52**(3/4), 335–377.
6. Eco, U. (1984). *Semiotics and the Philosophy of Language*, Macmillan, London.
7. Barthes, R. (1957). *Mythologies*, Seuil, Paris.
8. Feyerabend, P. (1975). *Against Method*, New Left Books, London.
9. Kuhn, T. S. (1970). *The Structure of Scientific Revolutions*, Chicago University Press, Chicago.
10. Clyde, W. M. (1934). *The Struggle for the Freedom of the Press*, Oxford University Press, London.

11. Febre, L., and Martin, H.-J. (1976). *The Coming of the Book*, New Left Books, London.
12. Marr, D. (1982). *Vision*, Freeman, San Francisco.
13. Yokota, M., Taniguchi, R., and Kawaguchi, E. (1984). 'Language–picture question–answering through common semantic representation and its application to the world of weather report'. In *Natural Language Communication with Pictorial Information Systems* (Ed. L. Bolc), pp. 203–255, Springer-Verlag, Berlin.

PART 4

Social Issues—Realities and Aspirations

Artificial Intelligence for Society
Edited by K. S. Gill
© 1986 John Wiley & Sons Ltd

13. NEW TECHNOLOGY AND THE DISABLED: WHAT RESEARCH IS USEFUL?

JOHN PICKERING Psychology Department and **GEOFF STEVENS**
School of Industrial and Business Studies, Warwick University,
Coventry

ABSTRACT

The paper looks at the potential effects of new technology on coping with disability. First, a brief overview of some aspects of disability is given. Next, the areas of work, learning and communication aids are used to suggest some applications for new technology. It is clear that new technology could have a great impact in all areas. What is less clear is how best to get the necessary research and development done. The last part of the paper offers some suggestions.

INTRODUCTION

No one doubts that new technology can have a beneficial impact on the lives of those with special needs—it is obvious. The real question is what sort of projects ought to be attempted and how such work is best brought about.

Accordingly, this paper offers some speculations on what might be done, both in the long and short term, to develop systems of practical use to the disabled. It also examines some issues in the research strategy needed to bring them about.

Throughout, the term 'new technology' has been used to cover the range of computer-related disciplines such as information technology, knowledge engineering and cognitive science. Presentation is highly informal and aims for breadth at the expense of detailed coverage. Accordingly no references are given though a short bibliography is provided.

There are three sections to the paper. The first deals with some aspects of disability, the second with what new technology might have to offer and the third with how support might be found in the research environment of the present and near future.

WHAT IS DISABILITY, WHO ARE THE DISABLED AND WHAT ARE THEIR NEEDS?

Disability is not a uniform state and those with special needs are a hugely diverse population ranging from those who have, for example, a sensory difficulty such as deafness or marginal vision but who are nevertheless capable of independent living to those with major physical and mental impairments and who are in long-term institutional care. There is a growing awareness among those who work with this population that the use of computers and information technology can help in a number of ways and much work is going on at present. It is, however, unrealistic to expect that such a diverse group can benefit uniformly—much needs to be done in matching needs to techniques. It is also unrealistic to expect that new technology, for all its power, is going to have any greater impact on the lives of the severely and multiply handicapped than good human factors engineering does at present. Nonetheless, for some groups of the disabled new Technology holds greater promise of liberation and independence than any other technology thus far.

How many disabled people there are in the UK population? A comprehensive survey of 1968–69 is still the only officially sponsored national survey apart from some information available from Census returns. There are more recent sources such as the reports of the various societies concerned with handicaps, and local authority records also provide more accurate pictures of the demography of disability, but there have been no large-scale attempts to assemble this information into a centralized resource.

There is nonetheless a fair degree of concensus on numbers. Some 3 million persons in the United Kingdom suffer from some sort of impairment and of these slightly less than half can be said to have a substantial handicap, i.e. one that is a major block to normal independent life. This figure is based on all types of handicap and makes no distinction between those who are institutionalized and those who are not.

One thing is clear from these figures, however approximate they might be. The disabled form, happily, a small proportion, around 2 to 3 per cent of the general population. Thus compared, say, to general health research applying to the total population the case for research support for projects on new technology and the disabled might do well to stress the particular suitability of new technology techniques as justification. More specifically, new technology offers something to this particular population, however small it might be, that cannot be done in any other way, thus justifying the need to devote resources to such work. The section on possible applications of new technology below will address itself to this point.

The sources of disability are varied. They range from congenital conditions such as spasticity and spina bifida to the after-effects of accidents and surgery. Whatever its source, whether a disability becomes a handicap depends upon the interaction between the disabled individual and their physical and social environment. Technology in general and new technology in particular is a major factor in this interaction and a few points about it are worth making here.

First, the nature and extent of disability is seldom, if ever, uniform. There are ranges of severity within a single handicap and multiple handicaps are frequent. Individual differences in training and compensatory skills make for great diversity in handicap profiles. Those working in the area are always quick to point out that each case is going to be different, and they are always right. Accordingly, there are, bound to be great variations in requirements, not all of which are likely to be matched by uniform application of some aspect of new technology.

Second, there are great differences between the effects of impairments acquired through illness or accident and those which are congenital or nearly so. This difference lies mainly in the level of competence reached in the fundamental skills of living and learning. In the case of computer-based work which can have a somewhat forbidding technological feel, this factor is likely to be all the more important in influencing the acceptability of such work by the disabled and their teacher/helpers.

Third, it has to be recognized, as pointed out above, that for a substantial proportion of the severely handicapped, particularly those handicapped from birth, the fundamental need is for help with everyday matters of independent living. Here new technology may have little to offer over and above that offered by skilled human factors engineering, at least at present.

Lastly, the socioeconomic context in which research into new technology applications in disability takes place needs to be borne in mind. In the information age, living and working may require less physical strength and mobility while computers and telecommunications will allow the handicapped to participate in society to a far greater extent than previously possible. However, computers could become a barrier to the handicapped in normal life unless special efforts are made to make computers far easier for them to use them is currently the case. Computers are being applied at an ever-increasing rate to amplify the abilities of the able-bodied, to increase their efficiency and productivity at work and to provide them with new opportunities. The education system is already computer aware, and employers are making more and more use of computers as they become more powerful, sophisticated and above all cheaper. Much work is being done on new technology applications to aid those with special needs, but there is a danger that it may focus on too-short-term objectives which are too limited. For example, there are now a number of microprocessor-controlled scanning/selection devices which mimic the action of the classic pre-new-technology aid, the Possum. They are well-designed, cheap and effective, but it is not desirable to equip those with special needs with yesterdays tools, however improved they might be, if the risk exists that the handicapped will be taking two technological steps forward while others are taking five.

Overall, then, while new technology holds great promise for help with a wide variety of special needs, it should not be forgotten that the population to which this help may be appropriate is not the whole of the spectrum of disability by any means. There are going to be strong constraints both on what can be delivered and also on the capacity for the user population to take it up. Also, projects should capitalize on just those aspects of new technology which offer the most liberating opportunities for the disabled as well as amplifying devices already in existence. Accordingly, the next section offers some examples, albeit highly speculative ones for the most part, to illuminate just what these aspects might be.

NEW TECHNOLOGY AIDS: POTENTIAL AND ACTUAL

This section offers some speculative applications of new technology in the design of assistive devices. It is not a review of work in this area though some examples given are of projects which are currently under way. Many of these projects seek to upgrade existing aids by incorporating new technology techniques. This effort is worth while and will occupy many researchers for some time, but this effort can be paralleled by projects which exploit the essence of the technology itself rather than use it to mimic devices already built in other ways. These sorts of projects are starting to appear and this section reviews some and suggests others. As a means of classifying projects the portmanteau usage of 'new technology' will be dropped and sections will deal with, respectively, information technology, knowledge engineering and cognitive science. The techniques and vocabulary of these areas overlap, but the division is useful as it makes a start on the exercise of matching special needs to the potentials of new technology.

Information technology

Here we deal less with particular computer-based aids or systems and more with how, say, work and learning might occur when current trends in the communications infrastructure of society become commonplace.

Work methods at all levels are becoming computer oriented to an unprecedented degree; remote working is starting to become a real possibility as the sophistication of remote workstations increases and as the costs of online access from home drops. Many activities in the commercial and industrial sector are now being done in entirely different ways, creating a new job market in which the disabled may be able to compete on more equal terms than before.

For this to happen more research and development is required in areas such as special keyboards, keyboard emulators, systems to allow special analogue input and voice-controlled equipment. The point is that if the handicapped are to be given an equal chance in the new work market, they must be able to gain access to and operate standard software. Employers will only be interested in whether a prospective employee can fit in with work systems already in place and whether they can work to acceptable rates and standards. Of course, despite the possibility of home-based work, there are still going to be problems of acceptability in relation to verbal and interpersonal skills since conventional contact with colleagues will not be eliminated by any means. Such a perennial problem is going to be solved directly by any of the projects suggested here. However, there may well be an indirect contribution, because if new technology allows disabled individuals to do jobs hitherto seen as being beyond their capacities, this will change their image for the better in a number of ways.

What then might be done to make remote working as easy for the disabled as it is for the rest of the working population? At least two strategies could be pursued, one being to provide the disabled user with special equipment, the other to modify the systems used. The former is to be preferred since it renders the disabled mobile in the job market and it is clear that making special requirements on employers will reduce the chances that they will offer such jobs to the disabled as readily.

Accordingly, a realistic project here is a home workstation specially designed to meet the needs of the disabled. The station would need to match the protocols and capacities of conventional equipment and also to match the needs and capacities of the user. No one design of workstation is likely to fit the bill in all circumstances and a possible strategy would be to assemble a toolkit of devices, a suitable selection from which would allow the majority of user/system requirements to be met. Such a toolkit will require such things as specially designed switches, keyboards and manipulanda in general, voice operation to permit the maximum amount of 'hands off' operation, flexible and powerful networking capacities, keyboard and analogue input emulators, user-configurable command libraries and much else besides.

Of course, as has ever been the case, a very good research strategy is to wait for someone else to do what you want, and the present case is no exception. Some components of the toolkit have been or will be developed in the commercial sector in any case and no special research and development need be directed at them; automatic dialling equipment would be an example. However, their incorporation into special needs workstations will require quite careful development work on compatibility. The recent work on standardizing input protocols for communication aid interfaces is an example of this sort of exercise. A substantial amount of work had to be done on this issue which is relatively simple when compared with the complexity of the communications engineering involved in remote access over public or private networks. Hence the problem of making sure any 'toolkit' can accommodate to the communications environment in which it will have to operate is not going to be a trivial one.

The provision of affordable and effective specialist workstations will not on its own do that much to improve the employability of the handicapped. Any such development on the hardware required to participate in the world of work and leisure that is being created by information technology must be accompanied by relevant training of users and alerting of employers to the employability of a group that they may not have considered before.

Also, we need to examine the roles created by information technology since not only are old jobs being done in new ways but also entirely new jobs are being created. Depending upon what these new roles require they may be more or less suitable to someone coping with the effects of a handicap. In any case, the information produced in such a survey would be of great use in designing training programmes. As an example, the computer training courses at a Further Education College for the Physically Handicapped in Coventry have stressed the acquisition of skills in the use of spreadsheet calculators rather than, say, general keyboard skills suitable for text processing or data entry. The rationale in doing so is that the rate and volume of keyboard work in the former skill would fall within the capacities of a larger section of the handicapped user population than in the latter.

In any case, the prospects for a more equitable share in the job market created by information technology for those with special needs are good. However, it needs to be borne in mind that the impact that information technology is going to make on the world of work is far from understood. It has often been suggested that far from creating new work and roles in which a living may be made, information technology will result in a net loss of employment. In some ways this has already happened, of course, in the shape of the huge volume of computer-based work which has replaced human skills in, say, the banking and insurance business. This, however, is not a disenfranchisement from the world of work that has caused much outcry because of the nature of the work replaced.

It is no great service to the disabled community to equip them with skills which serve only to commit them to the least attractive and rewarding of jobs. Rather, the aim should be to direct training at those roles created by information technology where the handicapped can function at a comparable level to the non-handicapped when suitably equipped and which also have some intrinsic worth.

The second area in which the implications of information technology are to be examined is learning. Here any number of examples, both actual and potential, of what might be done with this or that piece of information technology could be given since the possibilities for enriching the learning process are enormous. Rather than do this, however, what will be offered is an amplification of a point made briefly earlier using learning as an illustrative case.

The point was the strategic one concerning how the case for supporting new technology aids for the disabled might be made, namely that new technology offers not only ways of doing better what has been done before but also entirely new ways of matching the needs of some types of disability in a way that no other sort of techniques could.

In the case of learning, the disabled have both the needs of ordinary learners along with needs specially related to whatever handicap it is that they have. Information technology can help with both but it is in the second case that most interest lies.

As an example, take the problem of mobility and manipulation. Spatial learning, except when on the gross scale in a topic like geography, is expected to involve the learner in active exploration of the subject, be it by handling, dissection, construction/deconstruction or simply walking around. Piaget, Montessori and other contributors to the theory and practice of learning and teaching all stress the motor foundations of knowledge and the need for the active participation of the learner. The teaching of subjects like crafts, anatomy and some aspects of technical skills would not be expected to proceed without such active physical involvement of the learner.

While no amount of technology of whatever sort can fully remit some motor impairments, nonetheless advances in information technology allow the construction of compensatory aids or environments of a greater degree of sophistication than ever before. More to the present point, it is the nature of what information technology allows that is crucial, not just the fact that more assistance can be given than before.

Interactive video is an example of a technique which promises to be available at a realistic price within the next few years. What it offers in terms of high volume and quality graphics, learner-paced presentation of material and the like all address the needs common to all learners and as such will be powerful and usable. Over and above these sorts of use, interactive video may give some sections of the disabled population an opportunity for learning experiences simply unobtainable in any other way.

Take the present example of spatial learning. The familiarization of a learner with an environment or an object is typically a matter of the teacher providing the learner with the opportunity to investigate and leaving them to get on with it. For some kinds of motor disabilities this is not possible, but a useful approximation to it might be built using interactive video. Consider, for instance, the problem of getting to know an environment like a neighbourhood or large building. It is possible to construct an interactive video database about the environment which allows the user to explor it in an active way without actually being mobile within it. Similarly, some current CAD/CAM systems could allow an equivalent investigation of the spatial properties of objects without requiring conventional manipulative abilities.

These examples may, in terms of current development costs at any rate, be unrealistically expensive as serious contenders for research support. However, the principle they demonstrate is that new technology systems may provide unique opportunities to match special needs. Accordingly, this proposal may be of use in countering objections to supporting new technology work on the basis that it is, say, too expensive, too complex or directed at too small a population.

This section has considered only a few examples of how information technology might contribute to the education and employment of the disabled. The aim has been to look at what needs to be done to use some of the potential of information technology to improve work opportunities and to suggest one way to help such research get supported. This last point is not, of course, confined to information technology alone. It is just as relevant to the areas of new technology dealt with in the next two sections, namely knowledge engineering and cognitive science.

Knowledge engineering

This aspect of new technology involves such things as expert systems and intelligent databases among others. There are some special implications for the handicapped in the developments that are occurring or are likely to occur in the area over and above the changes that it will bring to work practices which have already been touched on. There is, for example, the possibility of designing supportive systems to provide a rehabilitation environment.

A significant section of the handicapped community are those who for one reason or another have acquired some impairment to their general intellectual and/or physical capacities and hence to any skills they might possess. The rehabilitation into an independent lifestyle requires a supportive environment, of which the most important parts are usually human helpers, trainers, assistants and the like. Someone who retains parts of a skill is still able to function and to regain competence so long as someone is available to help in the recovery of those lost parts of the skill.

There is fair evidence that the competence of expert systems and automated assistants could get to a level where this supportive environment could in part be provided by knowledge engineering techniques.

Take as examples the skills of electronic circuit design, fault finding in electromechanical machinery, designing office layouts and such legal skills as conveyancing and probate. All these are skills of a practical and valuable sort. They represent a selection from a rapdily increasing set of skills that are being realized in expert system form. Such systems function, for the present at any rate, not as replacements for human skills but as complements to it—as a supportive environment in which decisions are taken.

Were someone already possessed of life skills or in the process of acquiring them to become disabled, their position could be greatly helped by providing them with an environment where intelligent assistance was constantly available both for retraining and for exercise of residual skills. Such assistance will be most natural and effective when coming from a trained human helper, but there is fair evidence for the feasibility of supplementing this help with support systems based on knowledge engineering techniques.

This amounts to proposing that knowledge engineering could be used to build a skill backup system powerful enough to help individuals recover from the effects of illness or injury.

If this sounds improbable, it should be noted that for the past half decade or so there has been substantial investment in expert systems to capture, reproduce and extend a wide variety of intellectual, craft and professional skills. The expressed intention in many of these research projects has been to make the skill question resistant to the vagaries of the human population in which it exists. These vagaries are in fact the awkward tendencies of employees to retire, resign, ask for unacceptably high salaries and in other ways cease to deliver the needed skill.

If, as seems to be the case, knowledge engineering is taken sufficiently seriously as an adjunct to human skill, it seems quite possible that similar techniques could be used in supporting those with impairments to either exercising a skill previously learned or learning new ones.

Information technology concerns the transmission, storage and processing of information of whatever sort and deals to a large extent with the hardware and software that does that job. Knowledge engineering concerns information which represents the bodies of facts and techniques which are known as skills when possessed by human beings. Continuing this trend towards the specific from the general, cognitive science deals with the structures and processes underlying intelligent action of any sort and which therefore may be independent of any particular domain of knowledge. The next section considers possible applications of cognitive science to some aspects of disability.

Cognitive science

Cognitive science aims to formalize and hence to reproduce and explain natural intelligence, including that of human beings. Whether the new discipline will in fact do so need not concern us here since there are important enough implications for disability aids arising from the capacity to mimic quite simple aspects of natural intelligence. The term 'cognitive science system' will be used here to mean any computer system whose capacities derive from the use of the methods of cognitive science.

This section will look at just one of the needs noted in the introduction, namely that of communication. Here communication is used in a very broad sense to include not only the conventional meaning of conversations or message passing but also the process whereby someone transmits information with the intention of informing or controlling someone else or something else such as a computer system or a special environment. It will be suggested that here cognitive science could play a special role in assisting those with communicative difficulties through a process we shall call information amplification.

Handicap in communication is especially disturbing since it strikes most directly at human dignity and independence by rendering people powerless and isolated. It prevents the expression of a person's potential and stands in the way of the human contact that both generates and sustains a sense of self. The literature on communication handicap abounds with examples of handicapped individuals helped to communicate by teachers, parents and friends who learn to act as interpreters of whatever communicative repertoire the handicapped person has. This usually takes place over a long period of time and relies on intimate contact between the people concerned. Although the details of individual cases differ greatly, a common thread is that the restricted communicative acts of the handicapped individual are interpreted and expanded by the helper who has learned to recognize their meaning. This is information amplification at its most subtle and effective; it relies on sympathetic human contact and subtle interpretive skill only to be expected from natural intelligence.

Accordingly, it is not to be expected that anything like this level of effectiveness in information amplification could be got from cognitive science systems either at present or in the future. However, something like it, but much simpler, could still be of great importance in helping with communication disability, and here some possibilities might be suggested by the brief examples given below.

As a first example consider the idea of a script. This notion comes from the work on memory done by Roger Schank over the last decade or so. A script is a more or less stereotyped set of actions and interactions which describe the repeated events which occur in everyday life. Examples might be having a meal, either at home or in a restaurant, visiting a doctor, having a holiday or taking a bath. A script is a knowledge structure containing a description of the expected sequence of events making up the overall event pattern, reasons and consequences to do with these events, expectations about the actions of others in relation to the event, preconditions for actions and more.

Schank uses the script concept as part of his theory of human memory, though its value as a theoretical construct need not concern us in this context. One of the strengths of cognitive science, however, is the effort made to explain and objectify terms and concepts through the writing of computer programming. In the case of scripts, many illustrative programs exist mainly in the areas of modelling inference and understanding stories. Once again we need not be too concerned about how convincing or not these programs are; what is of interest is the possibility that they might be the means of importing some of what scripts can do into other areas.

Take, for example, the area of environmental control. The sorts of actions we all carry out within our environments conform to script-like patterns. Environmental control systems should aim to offer the disabled ways to alter their environment in accordance with what they wish to do with the least trouble. Were some form of script-like description of the preconditions and results of various actions to be incorporated into environmental control systems it could reduce the amount of communication that an individual needed to make in order to bring about the effects they wanted. For example, if a human helper of a disabled person was told that their charge wanted to make a 'phone call, certain inferences would be drawn about other requests implicit in the overall request, e.g. the 'phone would need to be available perhaps by being released from other equipment, the dialing system likewise would need to be readied perhaps by loading the necessary software, and so on. Rather than have to carry out these subtasks by a series of successive requests, the user need only state the top-level task request to have them carried out. Given sufficiently rich representations of the working environment a script-based controller would ease the burden on the user by making communication and control more natural.

Such information amplification occurs without effort in natural human discourse and again it is unrealistic to expect the same degree of naturalness from cognitive science systems, since the extent of the knowledge and the power of the processes which it would need are beyond the capacities of contemporary computer science. Whether this will remain so is another question. For the present, however, simpler forms of script-based information amplification tied to particular situations looks a feasible project for the near future.

The above is a proposal for a communication aid which can use stored knowledge to recognize and act upon the implications of a particular communication. A far more common form of information amplification which occurs repeatedly in the interaction between the communication impaired and their helpers is when the helper learns to recognize what a particular attempt at communication means *in toto*. Here, familiarity with who is doing the communicating and the context in which it is done are clearly the most relevant types of knowledge being used. Given the complexity of acquiring and representing such knowledge, it again seems unrealistic to expect cognitive science to be able to provide any sort of equivalently skilled system. At simpler levels, though, something may be possible.

Take as an example the construction of messages in natural language by someone who is both speech and motor impaired—a fairly common example of a communication disability. Many devices now exist to help with this situation and a lot of work is now under way to increase the range and sophistication of such devices. The anticipation of communicative intent referred to above is natural to this situation when seen from the hearer's point of view. When waiting for a message to be assembled by, say, the pressing of keys by a head pointer or by selection from some scanning menu device, the receiver of the message is often in the position of being able to anticipate what is being said and to save the sender effort by indicating that the sense of the current word or phrase has been understood.

Many different levels of knowledge are in use here, reflecting the variety of levels at which natural language is redundant. Some of this knowledge has to do with the semantics and pragmatics of language, which not surprisingly has been found to be extremely difficult to represent and use in even the most advanced computational linguistics. Other types of natural language redundancy, however, like the serial dependencies between letters and the syntactic constraints on words in sentences, may be representable in a sufficiently powerful communication aid. It may be possible, for example, to construct a system that would aid the process of writing by providing a dynamic menu of word choices as the character-by-character assembly of a passage of text occurs. To be useful, such a system would have to present a sufficiently large array of choices to stand a good chance of hitting the word actually intended but not too large an array as to need more time to scan than would be spent on typing the word anyway. Also, the content of the choice array needs to be based on the context of words typed, both syntactic and possibly semantic, as well as upon the characters typed at any particular instance.

Some of the requirements of such a system are large dictionaries with multiple-access criteria, some capacity for recognizing and using the constraints arising from natural syntax and powerful algorithms for the selection, ordering and presentation of a choice array. Some of these areas are the subject of current work in cognitive sciences and in computer science. Thought the power of the computers required to run such a system is prohibitive at present, this will not remain the case and accordingly such a project could well be undertaken now in anticipation that the results of any research would be implemented on the hardware that is likely to appear within the near future.

Of the three areas or levels of new technology, cognitive science provides the greatest scope for truly revolutionary aids. These will in effect be mimics of the intelligent help now only available from human helpers. This mimicry cannot be expected to be complete but it does not need to be of significance—a disability only becomes a handicap as a function of the environment in which the disabled individual lives. Surrounding the disabled with intelligent systems to assist in, among other activities, work, learning, environmental control and communication will contribute greatly to lessening the effects of some sorts of disability.

The three parts of this section have presented suggestions of what different aspects of new technology might have to offer those coping with disabilities. While the examples themselves are largely speculation, the pace at which new technology is developing makes it likely that what seems too ambitious or costly under present conditions tends to become feasible sooner than expected. This trend towards the accelerating power of computer-based devices will not on its own bring about the types of aids envisaged above. Other issues are involved, e.g. non-hardware limitations to new technology aids and the factors that will control the availability of research and development funding. The concluding sections of this paper look briefly at these issues.

NEW TECHNOLOGY: TOOLS AND SKILLS

If the pattern of advance in new technology hardware continues, we are heading towards a time when affordable powerful computer systems will become widely available. If we consider the computers of even the early 1970s and compare that with today's situation, a trend towards smaller, more independent but equivalently powerful computer systems is clear. The next decade and a half will surely provide a similar pattern.

The mere existence of this hardware base does not of itself mean that new technology aids like those envisaged here will necessarily be developed. The availability of hardware is in fact a minor consideration when compared with what else is needed for such work to get done. The most important is of course money. This will be looked at in the next section, while this section will consider other factors in the development environment, particularly skills, software tools and personnel.

For the application of new technology to aids for the disabled to occur, there will need to be developed a population of workers familiar both with the needs that are to be met and with the means of using new technology to do so. Work in cognitive science relies on a wide variety of skills and tools, particularly special programming languages, the development systems within which they are used and familiarity with a large repertoire of theoretical and practical ideas which form the basis on which large programming projects are built. What is going to have to happen is a transfer of technology, and this will take time. At the moment, the practice and products of new technology is concentrated in centres of higher education and in rapidly developing parts of the commercial and industrial sector. As such, the skills and personnel are not going to be available at the necessary level for some while yet. It will probably be the case that for some years there will be an excess of computing power over the research and development establishment able to make use of it.

As a particular example of this, consider software development tools. Such things as specialized languages, programming techniques and software development systems are now very much in the centre of research in computing science. The end result of this work will be software tools which will make the power of computers more accessible. Such tools and the skills to use them are important prerequisites for the kinds of development that have been proposed here, and a vital question is how they will become available to those working with the disabled. This could come about either by the recruitment of people familiar with the use of such techniques from other areas or by training of those already working within the field.

The sorts of computers and the software tools that exist at present are not yet powerful enough to support the kinds of new technology developments this paper has envisaged. In the near to medium future all the elements necessary to make it happen will become available and affordable. These elements are the hardware, the software tools and the personnel able to use them. The rate at which the first two elements appear is not under the control of those working in the field of disability. The third element, training of research personnel, is however, and efforts might well be undertaken now to provide training which will enable full use to be made of what new technology has and will increasingly have to offer.

There is likely to be a threshold effect in the application of new technology to assistive devices. Much of what would seem to be possible is nevertheless seen to be impracticable owing to the lack of sufficiently powerful techniques to bring it about. The rate at which deliverable computing power is increasing makes it likely that in the fairly near future a point will be reached where a truly intelligent assistive device will be designed, built and found to work satisfactorily. At this point it is likely that the practicality of such an approach will become increasingly clear and will necessarily come at a time when the hardware and software environment has developed to become more widely accessible and the rate of work on this approach to helping with disability can be expected to accelerate.

All this may seem wishful thinking in view of the fundamental question of cost. All the techniques and projects considered here in the end will involve substantial software development work and it is a commonplace in the computer industry that in the future the cost of developing software will increase enormously compared to the cost of the hardware on which to run it. Accordingly, some consideration should be given to how the sorts of work envisaged here might be brought about.

SUPPORT FOR NEW TECHNOLOGY

At present, research into disability aids and disability more generally is funded in ways which make it unlikely the sorts of new technology projects outlined here would attract much support. The special societies which concern themselves with one sort of disability are concerned that their funds be used in ways that have more immediate benefits. Research, however promising, that is not going to produce usable devices on a relatively short timescale is not going to be supported from these sources. This is not due to any lack of interest or sympathy with the approach on their part, but merely reflects the fact that such an approach does not fit their terms of reference. If the threshold for change in research and development suggested above does come about, then there may be a matching change in the funding policies of such bodies, as new technology aids will then be seen to be deliverable in a way that is not possible at present.

At a more local level, those working in special schools and colleges are not likely to see much of value in the investment of their limited time and resources in following new technology techniques at present. They, naturally, wish to see more immediate effects of what they do since their responsibilities to their students requires this. However, it is vital that such people are involved in the development of new technology aids as early as possible, since it is from them that the best practical evaluation of assistive devices can come. Some means must be found for working in a sufficiently well-equipped environment with personnel who have both the skills to work within it and who also have enough experience in or contact with the practicalities of work with the disabled to make what they do realistic and usable.

In the United Kingdom at present, the universities and polytechnics are still able to provide such an environment and, in fact, the current political pressures on them may well make the next few years an appropriate time for the development of projects of the sort suggested here. The pressure is for research in science and technology that is applied and which has direct social impact. Whether the sorts of projects proposed here exactly fit this bill is a matter of debate, but they clearly match to some extent.

Of course, proposing to site research in one environment or another has little to say about how it is to be funded. The fluid situation in the financing of university and polytechnic research work is likely to obtain for the next half decade, if not longer, and may again contain some unique opportunities. In particular, two factors in research funding policy might be considered.

The first is the stress on interdisciplinary research. The new technology aids envisaged here will require inputs from a variety of disciplines for their development, among which would be engineering, computer science and the various branches of psychology including ergonomics and cognitive science. The cross-disciplinary nature of such work combined with the social value would be strong factors in favour of it being supported by the research councils and by the research funds distributable by the universities themselves.

A second factor is the emerging pattern of cooperative support between the research councils and industry. While this support is mainly for areas where industry has a direct interest there could be a case made for new technology aids research fitting into such a pattern. At one level, the design and building of such aids is a special case of the work on man–machine interaction which currently is attracting substantial funding, and systems developed to help those with special needs are certain to be applicable to the rest of the user population. Also, there is increasing concern within industry that the social responsibility of companies be brought to the public notice. Whether this is merely a public relations exercise is an open question but it may well lead to genuine support similar to that offered by the oil companies to research into alternative energy sources.

Lastly, we may consider the recent moves on the part of local authorities to regenerate inner city areas through the development of small enterprises. Over the past few years a number of larger UK cities have explored with universities the possibility of cooperative ventures, often based on science parks, which exploit the research facilities of universities with the aim of setting up small concerns in areas close to where the university is sited. The sort of work proposed here could well be carried out in the form of a consultancy agency for the development of products whose building and marketing then became the concern of a local small firm established for the purpose.

The sources of support suggested above represent trends in the current political and economic climate from which research into new technology may be able to capitalize. As the accessibility of new technology techniques and tools grows so too will the recognition of its potential in the building of assistive devices. What this paper has sought to do is to speculate on what such aids might be like, to suggest ways in which a case for supporting the research necessary for their development might be made and in conjunction with this to take a brief look at the funding strategies that might be pursued.

The examples given may turn out to be too complex or impractical for other reasons, but even if the particular aids described here prove unrealistic there are clearly many possibilities of exploiting new technology in other ways. The time for this exploitation to become a practical reality is still some way in the future, but there is every reason to begin to prepare for it now.

BIBLIOGRAPHY

Articles and books

Hope, M. (1980). 'How can microcomputers help?' *Special Education: Forward Trends*, **7**(4), 14–16.
Hudson, E. (1984). *The Computer as an Aid to Those with Special Needs: International Conference Proceedings*, Sheffield City Polytechnic, Sheffield.
Rostron, A., and Sewell, D. (1984). *Microtechnology in Special Education: Aids to Teaching and Learning*, Croom Helm.
Schank, R. (1982). *Dynamic Memory*, Cambridge University Press.

Reports

A number of short reports on new technology and the disabled are available from the Department of Trade and Industry (Information Technology Division), 29 Bressenden Place, London SW1E 5DT.

A longer report (also available in summary form) entitled *Technology and Handicapped People* is available from the Office of Technology Assessment, United States Congress, Washington DC, USA 20510.

Artificial Intelligence for Society
Edited by K. S. Gill
© 1986 John Wiley & Sons Ltd

14. EDUCATIONAL ADAPTATION OF ETHNIC MINORITY PUPILS: RELEVANCE OF NEW TECHNOLOGY

GAJENDERA K. VERMA Department of Education, University of
Manchester

In the prevailing technological environment when computers are increasingly being introduced at all levels of education, it is most important that the design of new technological tools for education should not only focus on the educational and training needs of pupils but should also take into account social and cultural contexts within which those needs arise. If new technology is to benefit all pupils within the educational system, then it should consider, as a matter of urgency, how it can be used to fulfil some of the educational needs of ethnic minority pupils who are disadvantaged and face discrimination both inside and outside the school. In this paper I focus on the educational and occupational needs of ethnic minority pupils within the broader context of a plural society.

In Western countries there has been fairly explicit concern in recent years about the multiracial, multicultural and multilingual composition of society. This type of pluralism has arisen as a result of the demographic, social and cultural changes brought about by the process of post-war migration. This has generated discussion about the nature of society.

A major part of the debate has centred around education. Few would disagree that education is of crucial importance in an individual's life. Without the skills good education can provide the individual cannot hope to aspire to a job of his/her liking or perhaps to any job in the current state of economic recession and rising unemployment around the world. In Britain, the prospect of temporary and indeed permanent unemployment are becoming woven into the fabric of many young people's everyday experience. Although many ethnic and social groups are vulnerable, the brunt of unemployment has fallen particularly heavily on black and Asian youth, and unemployment among these groups is much greater than their white counterparts.

However, the school as a key institution has a major role in promoting individual self-employment and equipping individuals to meet the needs and expectations of society. It might be easier to attain these twin aims if society were homogeneous. In a plural society the population is composed of many people from varying socioeconomic, cultural, religious and linguistic backgrounds, and this diversity obviously affects the process of education.

Turning to the British context, in the last two decades considerable attention has been focused on the education of ethnic minority groups. At the initial stages of migration (during the 1960s) provision for ethnic minority pupils in the classroom consisted primarily of teaching them English. The fundamental philosophy of that time was the smooth absorption of ethnic minority pupils into the mainstream educational system. The 'culture' of ethnic minorities was regarded as a handicap to the process of schooling. By the 1970s the members of ethnic minority pupils in the classroom increased which changed the characteristics of the school population. Because of the patterns of immigrant settlement, i.e. concentrations mainly in urban areas, ethnic minority pupils became a distinct feature of particular schools. Issues then began to shift away from simply coping with life in British society to the more fundamental structures and ethos of the educational system. The questions were raised whether the educational system offered these pupils the same chances of academic success as those sought by the mainstream culture.

Issues like these in addition to questions regarding the ethnocentric stance of the school curriculum have become the focus of discussions. The nature of teaching materials, attitudes, teaching methods, testing and assessment procedures all embody ethnocentric overtones to a lesser or greater degree. Schooling also reflects values and ideologies of the dominant culture in society, and any educational programmes adopted by schools reflect those beliefs, values and attitudes that prevail in society as a whole. Studies have shown that disadvantages experienced by individuals in early stages of schooling can greatly affect their success at the later stages of education.[1,2] Research has also shown that blacks and Asians suffer from multiple, imposed disadvantage[3,4] as well as restricted employment. More recently Stone[5] has argued that it is issues of access to power and resources that are at the base of cultural groups, not poor self-concept as suggested by some researchers. This implies that it is poor schooling which has to be blamed for many of the problems encountered by ethnic minority pupils in British schools.

Research[6] has shown that ethnic minority pupils seem to experience considerable deprivation, which one would expect to have adverse effects on self-perception, motivation and educational performance. It should be noted that first generation young ethnic minority pupils had higher self-esteem than either the second generation (born in Britain) or their white counterparts. This may be explained in terms of mutual support between the first generation ethnic minority communities and the main cultural group. The second generation ethnic minorities not only perceive difficulties in getting employment but encounter prejudice and discrimination in entering both worlds of work and higher education. Their experience is likely to have affected their self-esteem. Many of these issues concerning the education of ethnic minority pupils have been highlighted in the Swan Report[7] which unfortunately have received scant attention by commentators of the Report.

In our various writings on education of ethnic minority pupils we have argued that bias in teacher perceptions of ethnic minority pupils are a major problem for educational practices, and imply the need for important changes in teacher training programmes.[8,9] If teachers, through pre-service and in-service training, can come to terms with the impact of prejudice and discrimination against ethnic minorities, then they may begin to mediate between pupils of different cultures, linguistic and religious backgrounds to foster tolerance, to fight racialism.

Teacher prejudice takes various forms. There is a good deal of evidence in the literature that in their ordinary classroom interactions teachers often operate on the basis of stereotypes, unproven assumptions and self-fulfilling prophecies. Such stereotypes are particularly strong in the teaching of ethnic minority pupils. Research by Haynes showed that teacher attitudes were likely to have a significant impact on educational performance of ethnic minority pupils.[10] Our research[6] also found that many teachers hold negative stereotypes of black and Asian pupils. Some teachers have a conventional view of ethnic minority pupils and hence they interpret all their problems in terms of their language difficulties, socioeconomic background and lack of intelligence. It is fair to say that views of many teachers are not only indicative of intercultural ignorance, but of ethnocentric rigidity.

Addiction to the assimilational position has led many teachers to ignore the reality of cultural, social and personal aspirations of ethnic minority pupils. In ignoring this identity the message from many teachers to many ethnic minority pupils, in the words of Jackson is: 'You do not exist'.[11]

Informal evidence also shows that the minority of teachers do express racist views in various social contexts. As Milner reported: 'I may have been too frequently horrified by staffroom conversations to feel that this minority is unimportant'.[12] Such sentiment was expressed ten years ago!

In the 1980s it is assumed that teachers are less likely to ignore the issue of ethnicity so far as the formal curriculum is concerned. Nevertheless bias in teacher perception of ethnic minority pupils is a major problem for educational practice. Elaborate curriculum models can be set out, courses and materials designed, which recognize, enhance and accommodate their social, cultural and personal aspirations. All will, however, come to nothing if teachers are not convinced of the desirability of trying to offer opportunity to ethnic minority pupils of whatever social class.

Teachers need to develop a multicultural, antiracist orientation so that they can act successfully and comfortably in a plural society. They need to learn how to acknowledge and, if necessary, reconstruct their own attitudes and practice towards, and assumptions about, those from a culture different from their own.

The unequal experience of black and Asian young people have become evident in many studies. Our research[6] showed that the careers element of the curriculum is one which receives scant attention in many schools and seems to have little impact on some pupils, particularly ethnic minorities, because of the incidental nature of such teaching. The careers advice that does exist in British secondary schools is not only inadequate but often biased as well.

Society and its institutions are not static. The evidence suggests that cultural pluralism in various forms is a becoming permanent feature of British society and is the single greatest development since the Second World War. In adapting to these changes, teachers and careers and education advisors could play a significant part in avoiding prejudicial treatment of ethnic minorities. Failure on their part would result in continued deprivation and disadvantage of future generations. The Swan Report has also shown that the disadvantaged position of ethnic minority pupils within the British educational system is due partly to ill-informed, prejudicial and inappropriate advice of career education.

The differentiation in cultural norms between ethnic minorities and the ethnic majority can also create a different set of problems in the education of minority pupils. For example, the experience and interpretation which an immigrant family develops concerning the mainstream community when their children come home after exposure to the British education system may be depressing; interactively, the socialization at home may come into conflict with values, beliefs and attitudes demanded at school. In this case, it is the young people who are most likely to encounter problems as they move between two cultures, and which contribute to their alienation.

Because of the practice of streaming in schools there is a progressive decline in achievement in some black and Asian pupils as they advance, increasingly alienated, through the school. The location of ethnic minority pupils in the educational hierarchy was demonstrated throughout the schools we studied.[6] We found that a large majority of blacks and Asians were placed in CSE (non-scholastic examination) and non-academic classes. There is a good deal of evidence from other studies that children of New Commonwealth origin are discriminated against in the British educational system. The outcome of prejudice and discrimination has been that disproportional numbers of ethnic minority pupils are assigned both to schools for the so-called educationally subnormal and are placed in lower streams of secondary schools.[13]

There is no shortage of evidence in the literature that existing testing and assessment procedures are biased against certain ethnic, cultural and social groups, providing an unreliable estimate of their achievement. Given the crucial issues and shortcoming associated with testing and assessment, there is a strong case for saying that the whole process is in need of careful evaluation and revision. In Britian, even today, many LEAs still make wide use of psychological tests and assessment procedures. Although standardized testing for selection purposes has been declining in the last ten years, testing and screening, monitoring, transfer and branding/streaming has increased. Since there are ethnic, cultural and class biases in the development and use of conventional tests, educational performance based on such techniques are of questionable validity.

In any discussion of the issues associated with differential achievement of pupils it has been consistently shown that an inappropriate and biased curriculum plays a major part in underachievement of pupils, particularly those from ethnic minority groups.[14,15] Some writers have focused on the curriculum as the primary vehicle for attaining equality of educational opportunity and consequently equality of outcome.

It is clear from the foregoing discussion that without coming to grips with the key issues of social and institutional change, mere tinkering with isolated aspects of education will have little impact. In brief, no real change will occur until the whole educational system changes. It will have to adopt an increasingly plural approach. The broad aim of education is to create a situation in which pupils of all races, religious and cultural backgrounds have an equal place in a harmonious, multiracial society.

If the educational system is to meet the cultural, cognitive and self-esteem needs of groups and individuals from diverse linguistic, social and ethnic backgrounds, the new technology could make a significant contribution to this end. To overcome many of the social and educational disadvantages of pupils, particularly those from ethnic minorities, an individualized structure of teaching based on new technology might prove useful. The new technology which emphasizes individual aspects of learning can be valuable. This approach must focus on individual needs and aspirations in a positive and non-stereotyped way. The thrust of my paper is that there must be individualization of teaching and a methodological individualism which recognizes and enhances the intrinsic aspirations of each pupil. Teachers should be aware of their personal cognitive styles and frames of emotional and value reference. They should be trained to utilize their cultural and cognitive understanding constructively in relating to the individual needs of each pupil.

I hope that issues raised in this paper would contribute to the widening of the debate on the relevance of new technology to the educational and training needs of our plural society. In the context of ethnic minority pupils, it is important to realize that their needs are also intertwined with their teachers' needs of understanding the plural nature of society. Recent developments in interactive video technology and artificial intelligence provide a powerful medium for developing interactive educational systems both for pupils and teachers. Such systems may use video to represent social and cultural situations within which learning becomes meaningful and relevant. AI techniques may be used to design systems which take into account individual needs of both the pupil and the teacher. Such developments have already been initiated at the SEAKE Centre which is concerned with the application of new technology for the education and training of disadvantaged groups. There is urgent need to make new technology relevant to the needs of all pupils and all teachers. However, to achieve this, it requires commitment and support from government and other agencies entrusted with the responsibility of education and training as well as those involved in AI and education research.

REFERENCES

1. Wedge, P., and Posser, H. (1973). *Born To Fail*. Arrow Books, London.
2. Rutter, M., *et al*. (1979) *Fifteen Thousand Hours*, Open Books, London.
3. Commission for Racial Equality (1978). *Looking for Work: Black and White School Leavers in Lewisham*, HMSO, London.
4. Lashley, H. (1981). 'Culture, education and children of West Indian background'. In *Teaching in the Multicultural School* (Ed. J. Lynch), Wardlock Educational, London.
5. Stone, M. (1981). *The Education of the Black Child in Britain*, Fontana, London.
6. Verma, G. K., and Ashworth, B. (1985). *Ethnicity and Educational Achievement in British Schools*, Macmillan, London.
7. Department of Education and Science (1985). *Education for All (Swan Report)*, HMSO, London.
8. Verma, G. K., and Bagley, C. (1979). *Race, Education and Identity*, Macmillan, London.
9. Verma, G. K., and Bagley, C. (1982). *Self-Concept, Achievement and Multicultural Education*, Macmillan, London.
10. Haynes, J. M. (1971). *Education Assessment of Immigrant Pupils*, NFER, Slough.
11. Jackson, B. (1979). *Starting School*, Croom Helms, London.
12. Milner, D. (1975). *Children and Race*, Penguin, Harmondsworth.
13. Tomlinson, S. (1980) *Educational Sub-normality — a Study in Decision-making*. Routledge and Kegan Paul, London.
14. Jeffcoate, R. (1979). *Positive Image towards a Multicultural Curriculum*, Batsford Academic and Educational Ltd, London.

15. Lynch J. (1983). *The Multicultural Curriculum*, Batsford Academic and Educational Ltd, London.

Artificial Intelligence for Society
Edited by K. S. Gill
© 1986 John Wiley & Sons Ltd

15.   THE EFFECT OF IT ON WOMEN'S LIVES

URSULA HUWS Freelance writer and journalist

This paper will consider the impact of IT on all aspects of women's lives: shifts in the boundary between domestic life and employment; changes in the structure of employment; changes in the location of employment; changes in the skill levels involved; and women's control over their work environment and daily lives.

I have been asked to take a very broad look at the effects of information technology on the totality of women's lives. This is an enormous subject—a sort of trot round the universe in 20 minutes. I do not propose to go into every detail; it would be absolutely impossible. However, I would like to preface what I have to say with a few general remarks because there are certain quite basic points that need to be made. A lot of people do not really see that different groups in society are potentially affected quite differently by information technology. Obviously IT does not have anything built into it which can tell what colour its users are or whether or not they have breasts or beards. You might say that in this regard the technology itself is neutral. However, different groups in society are differently affected by it and we have to understand why. Rather than go into the very complex reasons in detail, I would just like to draw your attention to two very obvious things which I think do illustrate my point.

One is the very different education and training which different groups tend to get in our society. Although girls and boys are taught nominally the same subjects at school, we all know that they generally emerge with very different qualifications; with girls overwhelmingly directed by all sorts of almost invisible pressures in the direction of arts and subjects like business studies or commercial studies and with boys tending to come out of school with the scientific qualifications, technical qualifications, and so on. Race is also a factor. We have seen from the education statistics that people from ethnic minority groups do tend to underachieve enormously in our education system.

People are emerging from the education system into adult life with very different degrees of confidence in the handling of technology and with different sets of skills, a fact which has to be borne in mind when we look at what happens to them later.

We are also talking about a society which is highly segregated in relation to occupation. Women officially form around 40 per cent of the workforce, and a higher proportion if the so called black economy is included. However, they are segregated into quite a small range of occupations. In fact over a quarter of women in employment are carrying out jobs which are over 90 per cent female, and a half of women in employment are carrying out jobs which are over 50 per cent female. Most men are working in occupations which are predominantly male while most women are working in occupations which are almost entirely female. Only 25 per cent of women are working in mixed occupations, e.g. social work, teaching, and so on. Women's jobs tend to be in the service sector. In manufacturing industry there are only really three industries where women are to be found in any numbers. These are the clothing and textiles industries, the electronics industry (doing the unskilled and semiskilled assembly jobs) and in food processing. Otherwise in manufacturing industries and extractive industries like mining and quarrying, the work is almost entirely male. Most women are working in service occupations: clerical work, nursing, school teaching, waitressing, cleaning, retailing, and so on. Thus there are huge concentrations of women in these industries. In fact 40 per cent of all working women are in clerical jobs. Numerically that is enormous; it is well over 4 million women in the United Kingdom working in jobs involving information processing.

When we talk about a technology which affects employment, we are actually talking about different sorts of jobs being affected quite differently. Obviously, no technology is going to affect all jobs equally. If a technology affects a particular type of employment, then it is going to affect different groups differently according to what sort of a job it is they do. To hypothesize, if the technology changed the nature of manual jobs in the service sector like cleaning, then it would disproportionately affect ethnic minority women. If it affected management jobs then it would disproportionately affect white men. We are not talking about a universal affect. This also has to be borne in mind when looking at occupational statistics. You very often get politicians arguing that x number of jobs have disappeared in one industry but that will not cause unemployment because a similar number of jobs have reappeared in another industry. In fact, however, if you look a bit closer it does not necessarily work like that. You do not find ex-coal miners working at supermarket checkouts. You are actually talking about jobs being created for a different group of workers than that which has been made redundant. I think that this is something which is very often ignored.

When we look at the latest wave of IT the jobs which have been affected up to now have been disproportionately female, because it is the service sector employment, particularly office work and retailing, that have been primarily affected by it. Up to the mid-1970s, these were the great labour-intensive areas of the economy, the areas that had not already been mechanized or automated in previous waves of technical innovation. I would like now to look at what has actually been happening to various jobs, and also what has been happening to areas of labour which are unpaid, which of course women are also responsible for. It is women who carry out most of the unpaid work in our society, the consumption work: going to the shop, housework, looking after children, and so on.

What has happened already as a result of IT and what is likely to happen in the future? I find that the most useful way of looking at what is happening to employment as a result of IT is to see it effectively as a two-stage process, bearing in mind that some industries are in stage two while other industries are only in stage one and it is not an even development. It seems to me that one can identify two broad stages and the first stage is computerization. It consists of the introduction of a whole lot of small one-off computer systems, perhaps a word processor in an office or an automated checkout in a shop. Stage two is the linking of them all together, which is not so much hardware based like the first stage as software and telecommunications based. We are already in a position to see quite clearly what computerization has done to a whole range of jobs which were not computerized before the late 1970s, but we are only at the beginning of seeing what the second stage does to jobs, because we do not yet have a very good telecommunications infrastructure which enables all these separate one-off computers to be linked up to each other in a fully interactive, efficient and economically viable way. That does seem, however, to be round the corner.

I will talk a bit about stage one first and then speculate a bit about what stage two means, remembering always that the two cannot entirely be separated, and there is inevitably some overlap between them.

The first thing which has happened has been that job loss has taken place in some industries. That has not actually happened on the enormous scale that was predicted in the late 1970s. My belief is that it is not until stage two is in operation that such a loss would really happen, because most information processing jobs still involve moving information from a bit of paper into an electronic system and back onto paper again. It is shuffling all those bits of paper around that really constitutes the bulk of what an awful lot of office workers are doing. Information systems are mostly still paper based, although some of the processing takes electronic form, and it is only when we move away from paper that the real job loss will take place.

We have also seen skill changes in a whole variety of jobs. I do not like using the word 'deskilling' which is popularly used in this context because I think the whole notion of how you define skill is actually quite complex. I think a lot of jobs are defined as 'skilled' simply because the people who have carried them out in the past have been in a position to organize, to ensure that there is limited access to those skills and knowledge. A lot of jobs which are defined as unskilled are in my view a lot more difficult to perform than other jobs which are defined as skilled.

The skill changes which have taken place as a result of computerization are actually quite contradictory in their effects in relation to gender. Many jobs which have been traditionally defined as highly skilled and carried out exclusively by white men have become 'deskilled'—printing being a very good example. Other instances can be found in the engineering industry and among others in the clothing industry. Such jobs used to have a long apprenticeship and the men who did them were the elite of their trades. Now they have been simplified, casualized and opened up for women and for people from ethnic minority groups and others who were excluded from them in the past. You could say that this has created new opportunities for employment for women, but it is also possible to argue that there has simply been a reduction to jobs.

There have also been enormous changes in the structure of jobs which have traditionally been done by women. For instance, secretaries in offices whose skills were based largely on shorthand and typing now do not need the shorthand at all. They need word processing skills which are very different because mistakes do not matter in the same way that they used to. There have also been changes in the organization of work. Jobs in many offices have become much more specialized and people have tended to become just a keyboard operator or just a filing clerk or just somebody who answers the telephone, instead of a mixture of all these things which is what was done before. This is because the process of automating a job involves analysing it and breaking it down into its component parts, which tends to lead towards that sort of rationalization. Indeed, one American survey estimated that of the productivity change that came about from introducing word processing into an office, 90 per cent of the productivity increase was actually due to the reorganization and only 10 per cent was due to the actual capability of the machines to produce greater productivity.

Another thing which has made a very great difference to management systems within offices and shops and so on has been the fact that IT enables a much tighter monitoring and control of the workforce. IT makes it possible, for example, to count how many key strokes per minute an operator is performing, what her error rate is, and when breaks are taken. This has had several effects. It has reduced the number of supervising posts which were traditionally the promotional path up from what you might call the shopfloor of the office. It has also been one of the factors that increases stress. In addition, it can provide much more sophisticated management information which has enabled decisions to be made much more quickly than before. It has also been a factor in some of the other changes which I will go on to talk about later.

There are some other effects which are rather less well studied. One of these is to encourage the externalization of labour. One of the main savings to companies introducing IT is not from getting rid altogether of labour carried out within their organization, but from transfering it to some other organization. In the case of the travel industry, for instance, tour operators used to have to keep an enormous staff to deal with bookings, and enquiries about the availability of bookings. That has now changed. By installing view-data terminals in the offices of travel agents, they have been able to transfer all that clerical labour onto the travel agent and cut their head office staff down considerably. A very similar process has taken place in some insurance companies which have transferred work onto the insurance brokers in the same way. In car part supplies, it is now in the garage that they actually check the availability of parts and so on, on remote terminals, rather than at the supplier's office.

This is not actually anything terribly new because if we look at the next stage of that process we see that it is possible to cut out the agent altogether, and transfer the labour directly onto the consumer. Looking back a few decades, some of us can remember the time when shop assistant jobs consisted of taking things down from the shelf and packing them and having time available to hang around and wait to serve customers. There were delivery boys transporting goods to people's homes. All that is now done by the consumer in unpaid time and IT has enabled similar forms of self-service to be developed in a whole range of other areas, too, with banking being a very classic example.

Automatic cash-dispensing machines have already had a very dramatic effect on staffing levels in banking, although banking is still expanding as a sector in terms of the numbers of accounts dealt with and so on. A lot of hope has been pinned on banking and insurance as a source of new jobs. If you look at all the structure plans that were produced at the end of the 1970s they all imply that employment levels will increase because finance and insurance are going to expand and provide many new jobs. In fact jobs are now starting to disappear in both banks and insurance companies. As a result of this, and other aspects of automation, the transfer of labour onto the consumer is taking place.

Not only do such displacements affect women's and men's jobs differently, depending on where in the labour market they are concentrated. They also disproportionately affect different groups in the community because some people are users of particular sorts of services more than others. Women are the main users of shops, and it is women who are the main users of hospitals, not only for their own health care but because it is women who tend to be responsible for taking sick children and elderly people and people with disabilities to hospitals. So even the introduction of things like diagnostic computers into the health service is likely to create more unpaid labour for some groups in the community than for others. This is a factor which cannot be quantified in neat financial terms, so it gets left out of a lot of equations. However, it does actually have quite an important consequence for the quality of life for people, as anyone knows who spends their Saturday mornings in a supermarket or regularly has to deal with out of order telephone boxes, launderettes or cashpoint machines.

So far I have discussed some of the things which are already happening. There are many other things which seem likely to happen in the future when we have the telecommunications infrastructure which I talked about earlier, with good, cheap, satellite links between countries and cable links or digitalized, upgraded telephone networks, within countries—when we effectively have an infrastructure that links one computer terminal easily with another or with a mainframe, things like electric fund transfer, home banking, and home shopping will be available. These developments are not only going to change the number of jobs in the sectors of the economy that are directly affected but they are also going to change the quality of life in the home. They are going to make distance a much less important factor in the organization of office work.

Telecommunication costs may soon become so cheap that they are effectively negligible in the way that computing power is now negligible for a whole lot of applications in offices. The cost of microcomputers is now so tiny compared with the salary of the average typist that it is not really an important factor to a management considering reorganization of an office. If we imagine a situation where telecommunications costs are negligible in the same way, and combine this with the fact that workers can be remotely monitored by machine, then we have a situation where office work, and indeed a lot of other sorts of work, can be carried out almost anywhere you can put a terminal. It is possible to see the beginnings of such a relocation already in the United States and to some extent in this country, where there has been a tremendous growth of a process known as 'suburbanization' in office employment.

San Francisco is one example of this. There downtown offices are becoming empty as companies open up branch offices in the suburbs with much lower overheads and with a fairly captive group of workers. These are mainly housewives drawn from neighbouring estates who find it more convenient to work in this way partly because the transport network is not particularly good but mainly because their time is at a premium because of the need to care for their families.

There is also relocation between one part of the country and another, which has happened in Britain also. For instance, some of the building societies have lately started having labour-intensive data-entry work and record keeping done in areas of high unemployment in the north of

England and sending the work back to London or Bristol by electronic means. This type of relocation is also happening in a more dramatic way with work being sent offshore from the United States using satellite links.

This takes place on a big scale in the West Indies and also in South East Asia. Even in communist China, there is now a free trade zone where the main activity carried out is data entry for updating American databases. They employ women working on extremely low rates of pay, and they have discovered that if they do not know English they work faster because they are not distracted by the content.

One could say that for this kind of labour-intensive data entry there is already effectively an international division of labour rather like the international division of labour that grew up in the manufacturing industry in the 1960s and 1970s. Offshore information processing takes place in many parts of the world—in the Philippines, in Brazil (one major American bank gets all its number work done in Brazil now), India, the West Indies and even in Ireland. It is my belief that this relocation of very labour-intensive work in huge chunks abroad is a transitional phenomenon, because I think within the next decade the technology is going to be there to make this large-scale manual data entry a thing of the past. With more sophisticated voice input and with better optical character recognition and so on, we are really not going to need that sort of data entry in bulk, especially when all the existing paper-based systems have already been computerized. However, at the moment offshore information processing is happening on quite a big scale and is a significant factor in the lowering of clerical wage rates in the United States.

In the long run, what may well become a more permanent feature is the shifting of many kinds of work to the home. At the moment in this country telecommunications costs are still relatively high for people doing this. You either have to put in a land-line which may cost well over a thousand pounds or you have to have the operator linked up on a modem using the British Telecom network, which is actually clocking up quite a lot of call charges over an extended period. Nevertheless, a number of people are already changing to electronic homeworkers; these are people whose labour is relatively expensive like computer professionals and executives.

The homeworkers I have come across working remotely have included some word processor operators, but have in the main been computer programmers, system analysts, technical authors, people working in the computer industry itself, all doing work that involves a certain amount of computer literacy. Needless to say they are almost all women with young children. They are also generally paid very much below the going rate for the job. It seems to me that this type of homeworking is likely to spread downwards as it already has in the States, as the telecommunications networks become cheaper, into many of the routine, clerical sorts of jobs.

This really does have pretty depressing implications for women. Homework is often presented as a very positive option. It is frequently said that one of the great advantages of IT is that it enables you to work in the comfort of your own living room. However most people's living rooms are not actually designed to give them space to work in any great comfort, and it is generally assumed that the advantages of doing this will be that you can do it and have young children around at the same time. In fact I recently read an article in a women's magazine which was about how to keep your children quiet while you work from home. The advice in it I found absolutely terrifying. It was completely Pavlovian. You were supposed to start working for ten minutes at a time, then build it up to half an hour, then an hour and so on. You gave the children a reward if they were quite and you penalized them if they were not. You were supposed to have achieved success when you had a child who would stay quiet for four hours while you worked. We know that to develop fully children need stimulation and they need care and they need attention, and for this to be presented as a positive way of looking after children seems to me to be absolutely awful.

We also know, of course, that homeworking is associated with many other problems: low pay, lack of trade union organization, lack of things like sick pay and holiday pay, pensions and other benefits. So this seems to be an area where we are going to get a lot of problems, particularly isolation.

This development is happening in the context of a general casualization of work, which is in itself penalizing people from ethnic minorities and women, because it is the kinds of jobs where they tend to be concentrated which mainly seem to be affected. It is very difficult to get accurate data on this because, although some employers do just close down functions in an office lock, stock and barrel and reemploy people working at a distance, in most cases it does not happen like that. The company may decide to start subcontracting work which previously was carried out by direct employees, for instance the data processing department or the advertising department, or whatever. Partly because automation has routinized these jobs and made them increasingly easily quantifiable, more and more are now being subcontracted in this way. It is at that point that a subcontracting agency comes in, either an individual already working from home or a small company employing homeworkers or employing people in a small off-site office.

One cannot monitor this as easily as one could if there were still an employer/employee relationship with the workers. Such subcontracting arrangements are associated with self-employment or with temporary short-term contracts. In fact office work seems to be moving towards the sort of employment structure that is usually associated with manufacturing industries like the clothing industry. In many ways the clothing industry provides us with a model for what is likely to happen in a whole lot of other industries as a result of IT.

In summary I do not think these effects are going to be all embracing; they are going to affect the quality of life in the home and outside the home. They are going to affect the nature of paid work and the nature of unpaid work. It is also, of course, important to talk about some of the positive effects these developments might bring, but these are obvious to any with a fertile imagination.

Note: This presentation was enlivened for the audience by the exuberant presence on the platform of the speaker's three-year-old daughter.

Artificial Intelligence for Society
Edited by K. S. Gill
© 1986 John Wiley & Sons Ltd

16.

LITERACY— COMPARATIVE PERSPECTIVES: 'AUTONOMOUS' AND 'IDEOLOGICAL' MODELS OF 'COMPUTER LITERACY'

BRIAN. V. STREET School of Social Sciences, University of Sussex, Brighton

Attempts to spread the uses of computing beyond specialist areas are currently being referred to as 'computer literacy'. The intention is clearly to draw an analogy with better known forms of literacy and to assert that what has happened there, in terms of mass literacy campaigns and attempts to break elite monopolies of the forms of reading and writing, should also be attempted with relation to computing. It is claimed that the population as a whole should be given access to at least the 'technical' skills involved in basic programming and key-board familiarity just as, it is assumed, formal education systems now provide access for all to the 'mysteries' of the written word. I would like to argue, however, that the analogy is not as simple as it seems, but rather raises a whole series of problems which have only recently been confronted in relation to literacy in general. Those wishing to spread computing to a wider public—and I would endorse their aims—need to recognize that it is not a simple matter of following the path of school literacies, as though they had forged the tools and solved the problems already. Computer teachers and organizers should not, in fact, expect to find a ready-made model in literacy practice, to which they can hitch their own aims and objectives and then wait for results. Rather, the whole issue of literacy in the conventional sense is currently being rethought and newly theorized in ways which challenge the everyday 'common sense' conceptions of it that I suspect are being assumed when the analogy is applied to computing. I would like to explore some of these recent debates within literacy studies and suggest that they are very relevant to the issues raised here, even though the terminology and assumptions may be unfamiliar to those directly involved in computing. To some extent I will leave it to readers to make many of the links for themselves, since I am not a computer 'buff' and would hesitate to treat too far into the field itself. The question of 'applications', though, seems to me one that requies some dialogue between the two fields of computer studies and literacy studies.

For present purposes I shall use the term 'literacy' as a shorthand for the social practices and conceptions of reading and writing. I have attempted elsewhere to describe some of the theoretical foundations for a description of such practices and conceptions and to challenge assumptions, whether implicit or explicit, that until recently dominated the field of literacy studies.[1] I contend that what the particular practices and concepts of reading and writing are for a given society depends upon the context—that they are already embedded in an ideology and cannot be isolated or treated as 'neutral' or merely 'technical'. What practices are taught and how they are imparted depends upon the nature of the social formation. The skills and concepts that accompany literacy acquisition, in whatever form, do not stem in some automatic way from the inherent qualities of literacy, as some authors would have us believe, but are aspects of a specific ideology. Faith in the power and qualities of literacy is itself socially learnt and is not an adequate tool with which to embark on a description of its practice.

Many representations of literacy, however—and, I suspect, of 'computer literacy'—rest on the assumption that it is a neutral technology that can be detached from specific social contexts. I shall argue that such claims, as well as the literacy practices they purport to describe, in fact derive from specific ideologies which are mostly not made explicit. This failure to analyse the ups and consequences of literacy and to theorize just what is the nature of the practice which is assumed to produce these 'uses and consequences', is particularly crucial when we come to consider comparisons of literacy practice across different cultures or across different cultural groups within a given society or nation state. Yet it is precisely such a 'cross-cultural' comparison that is occurring when members of a 'literate' group attempt to spread their practice to others—to 'non-literates' or 'illiterates' as they are referred to in this society, terms which already demonstrate the cultural loading and assumptions that underly statements about literacy. If we take literacy as referring to a set of conceptualizations and beliefs, rather than simply a technical skill, then it becomes clear that the imparting of this to others is a matter of 'culture transfer' rather than simply of 'technology transfer'. The implications of this for pedagogy and for policy are only just being recognized, groups like CAAAT at Brighton Polytechnic and, of course, SEAKE itself being still somewhat of pioneers in a field generally dominated by more 'technicist' models of computing and of literacy.

In order to clarify the differences between these different approaches to the analysis of literacy, I shall characterize them as the 'ideological' model and the 'autonomous' model of literacy respectively. I shall deal first with the 'autonomous' model. This model is often at least partially explicit in the academic literature, though it is more often implicit in that produced as part of practical literacy programmes. The model tends to be based on the 'essay-text' form of literacy and to generalize broadly from what is in fact a narrow, culture-specific literacy practice. The model assumes a single direction in which literacy development can be traced, and associates it with 'progress', 'civilization', individual liberty and social mobility. It attempts to distinguish literacy from schooling. It isolates literacy as an independent variable and then claims to be able to study its consequences. These consequences are classically represented in terms of economic 'take-off' or in terms of 'cognitive' skills. The main outlines of the model occur in similar form across a range of different writers both in the literacy field and, I would hazard a guess, in the field of 'computer literacy', although in the latter case I would be interested to hear counter-arguments or confirmation.

An influential example of the 'economic' case for literacy, for instance, and one which obviously has close links with assumptions currently being made about the implications of computing for society at large, is the claim by Anderson[2] that a society requires a 40 per cent literacy rate for economic literacy programme outlines. What is not specified is what specific literacy practices and concepts 40 per cent of the population are supposed to acquire. Yet comparative material, some of which Anderson himself provides, demonstrates that such practices and conceptions are very different from one culture to another. The homogenization of such variety, which is implied by the statistical measures and the economic reductionism of these approaches, fails to do justice to the complexity of the many different kinds of literacy practice prevalent in different cultures. It also tends implicitly to privilege and to generalize the writer's own conceptions and practices, as though these were what 'literacy' is.

The theory behind these particular conceptions and practices becomes apparent when we examine the claims made for the 'cognitive' consequences of literacy by many writers in the field. These claims, I argue, often lie beneath the explicit statistical and economic descriptions of literacy that are currently dominant in much of the development literature.

The claims are that literacy affects cognitive processes in some of the following ways: it facilitates 'empathy', 'abstract context-free thought', 'rationality', 'critical thought', 'post-operative' thought (in Piaget's usage), 'detachment' and the kinds of logical processes exemplified by syllogisms, formal language, elaborated code, etc. It would be interesting to explore how far similar assumptions are being made about the 'cognitive' consequences of computer literacy.

Against these assumptions I have posed an 'ideological' model of literacy.[1] Those who subscribe to this model concentrate on the specific social practices of reading and writing. They recognize the idealogical and therefore culturally embedded nature of such practices. The model stresses the significance of the socialization process in the construction of the meaning of literacy for participants, and is therefore concerned with the general social institutions through which this process takes place and not just the explicit 'educational' ones. It distinguishes claims for the consequences of literacy from its real significance for specific social groups. It treats sceptically claims by Western liberal educators for the 'openness', 'rationality' and 'critical awareness' of what they teach, and investigates the role of such teaching in social control and the hegemony of a ruling class. It concentrates on the overlap and interaction of oral and literate modes rather than stressing a 'great divide'.

The 'ideological' model has the following characteristics:

1. It assumes that the meaning of literacy depends upon the social institutions in which it is embedded.
2. Literacy can only be known to us in forms which already have political and ideological significance and it cannot, therefore, be helpfully separated from that significance and treated as though it were an 'autonomous' thing.
3. The particular practices of reading and writing that are taught in any context depend upon such aspects of social structure as stratification (such as where certain social groups may be taught only to read) and the role of educational institutions (such as in Graff's[3] example from nineteenth century Canada where they function as a form of social control).
4. The processes whereby reading and writing are learnt are what construct the meaning of it for particular practitioners.
5. We would probably more appropriately refer to 'literacies' than to any single 'literacy'.
6. Writers who tend towards this model and away from the 'autonomous' model recognize as problematic the relationship between the analysis of any 'autonomous', isolable qualities of literacy and the analysis of the ideological and political nature of literacy practice.

The writers and practitioners I am discussing do not necessarily couch their arguments in the terms I am adopting. Nevertheless, I maintain that the use of the term 'model' to describe their perspectives is helpful since it draws attention to the underlying coherence and relationship of ideas which, on the surface, might appear unconnected and haphazard. No one practitioner necessarily adopts all of the characteristics of any one model, but the use of the concept helps us to see what is entailed by adopting particular positions, to fill in gaps left by untheorized statements about literacy, and to adopt a broader perspective than is apparent in any one writer on literacy. The models serve in a sense as 'ideal types' to help clarify the significant lines of cleavage in the field of literacy studies and to provide a stimulus from which a more explicit theoretical foundation for descriptions of literacy practice and for cross-cultural comparison can be constructed.

The models are not logically equivalent, in that the 'ideological' model actually subsumes the assumptions implicit in what I term the 'autonomous' model. Those who subscribe to the 'autonomous' model are, in fact, being ideological, but in covert and often subconscious ways. They are, in fact, responsible for the polarity by their attempt to abstract supposedly 'neutral', 'technical' aspects of literacy from the cultural and ideological context, as though these could be considered independently and the 'cultural bits' added on later. The 'ideological' model, on the other hand, does not attempt to deny technical skills or cognitive aspects, but rather recognizes that they cannot be handled independently of the cultural whole which gives them meaning. It makes explicit, as far as possible, the specific ideological assumptions being made rather than denying that they exist in this context and thereby simply disguising them.

I use the term 'ideological' to describe this approach, rather than the less contentious or loaded terms 'cultural' or 'sociological', etc., in order to signal quite explicitly that literacy practices are located not only within cultural wholes but also within power structures—the very emphasis on the 'neutrality' and 'autonomy' of literacy by many writers is ideological in the sense of veiling this power dimension. I do not use the term in its old-fashioned Marxist (and current anti-Marxist) sense of 'false consciousness' and simple-minded dogma, but rather in the sense employed within contemporary studies of discourse (e.g. within cultural studies, social anthropology, sociolinguistics, etc.) where ideology is a site of tension between authority and power on the one hand and individual resistance and creativity on the other. This tension operates through the medium of a variety of cultural practices, including particularly language and—I would add—literacy. It could be argued that it has been lack of attention to this complexity of cultural practices and the tensions between them has led to the well-documented 'failure' of so many literacy campaigns.

At a recent conference on literacy, called precisely to analyse this failure and to suggest new conceptualizations and theories for handling the question that might be applied by Unesco and national agencies in the field of both adult and child education, a speaker suggested that the reason why governments continued to pour so much money and resources into programmes that had such a poor 'success' rate was not that they held overoptimistic assumptions about what literacy could do, which is currently a major argument in the field. Rather, he claimed, it was precisely because governments knew that literacy on its own could not achieve the aims laid out in the 'autonomous' model—economic take-off, social mobility, cognitive improvement, etc.—that they continued to give it such a high profile: it enabled them to avoid the real issues that such aims involve, issues of power and of the political relations between those in authority, with the power to define others and their 'needs', and the various competing groups and cultures in a given society or state. No one can easily stand up and argue that literacy campaigns are not necessary (my own position would be that some programmes are essential but that the way in which they are conducted is the crucial point), and so through their literacy programmes governments can give an appearance of appropriate action and good intentions that would be difficult to challenge, while at the same time maintaining and disguising the real power relations and forms of exploitation that create the problems they claim literacy can solve. This may be a somewhat cynical view for an English audience socialized into believing that governments are relatively well meaning and even, in matters of education and literacy, 'non-political', but it might help to put into perspective some of the policy decisions currently being made in the areas respectively of technology and education in this country. I am thinking in particular of the increase in resources for technological and computing 'hardware', which is what has helped to make 'computer literacy' a major issue in relation to both schooling and remedial education, and the decline in resources for the 'humanities' and social sciences in those same sectors. Questions about 'computer literacy' and how it can be widely disseminated can, it seems to me, only be confronted meaningfully after prior analysis and scrutiny of the bases of such policy decisions. These decisions arise out of deeper and often hidden ideological and cultural assumptions and it is to these that I would direct the attention of those concerned with computing and education, rather than to the technical and pedagogic 'problems' it raises. Within this society current debates about the 'spread' of computer literacy to working class groups and ethnic minorities are clearly related to differences of power and ideology between the various groups concerned, rather than to simply their relative 'educability' or cognitive competence. I am arguing, therefore, that it is both intellectually ill-founded and politically naive to attempt to conduct a discussion about 'computer literacy' in terms simply of technology transfer and pedagogic technique.

REFERENCES

1. Street, B. V. (1984). *Literacy in Theory and Practice*. Cambridge University Press.
2. Anderson, C. A. (1966). Literacy and schooling on the development threshold:

some historical cases, in Anderson, C. A., and Bowman, M. (eds.) *Education and Economic Development*, Frank Cass, London.
3. Graff, H. J. (1979). *The Literary Myth: Literary and Social Structure in the 19th Century City*. Academic Press, New York.

PART 5

AI, IT and Education

Artificial Intelligence for Society
Edited by K. S. Gill
© 1986 John Wiley & Sons Ltd

17.

IT, AI AND THE ELECTRONIC SABRE-TOOTH

DAVID SMITH Centre for Evaluation of Information Technology in Education, National Centre for Educational Research, Slough

PRELUDE

This paper draws its title from Harold Benjamin's brilliant parable 'The sabre-tooth curriculum'.[1] Readers will, I am sure, forgive me if I spend some time in drawing a lightning sketch of Benjamin's masterpiece. . . .

Once upon a time, in the paleolithic era, there was a tribe of caveman. Their life was simple, but hard. Then as now there were few lengths to which men would not go to avoid the pain and labour of thought. A thinker arose, nonetheless. His name was 'New-Fist-Hammer-Maker' (or 'New-Fist' for short). He began to catch glimpses of ways in which life might be made better for himself, his family and his tribe.

'If I could only get children to do the things that will give more and better food, shelter and security,' thought New-Fist, 'I would be helping the tribe to have a better life. . . .' Having set up an educational goal, New-Fist proceeded to construct a curriculum. 'What things must we tribesmen know in order to live with full bellies, warm backs and minds free from fear?' he asked.

New-Fist discovered three items which were central to the life of the tribe:

woolly-horse-clubbing-for-food
fish-catching-with-bare-hands
sabre-tooth-tiger-scaring-with-fire

He brought his own children up using his curriculum, and they thrived. Some of the more intelligent members of the tribe did as New-Fist had done, and the teaching of fish-grabbing, horse-clubbing and tiger-scaring came to be accepted as the heart of real education.

Gradually, being an educational statesman as well as an educational administrator and theorist, New-Fist overcame all opposition (practical, theoretical and theological) to his curriculum. Long after his death, the tribe prospered.

However, times changed. A new Ice-Age approached. Fish grew difficult to catch with bare hands (and all those stupid enough to be caught had long gone!). The woolly horses migrated away, to be replaced by agile antelopes. The sabre-tooth tigers died out, and their place was taken by cave bears which were afraid of nothing (and certainly not fire). The tribe was in danger of becoming extinct.

Fortunately for the tribe, there were men in it of the New-Fist breed, men who had the ability to do and the daring to think. Between them, they solved the tribes survival problems by inventing fishnet making, antelope snare construction and operation and cave-bear catching in pits. Once again the tribe prospered.

There were a thoughtful few who asked questions as they worked. Some of them even criticized the schools. They asked why these new subjects should not be taught. But the majority of the tribe had long ago learned that schools had nothing to do with real life, and the wise old men who controlled education had another answer: 'It wouldn't be education: it would be mere training, and anyway the curriculum is already too crowded. What our people need is a more thorough grounding in the basics.' The old subjects were taught not for themselves, but for the sake of generalized skills.

However, the radicals persisted: 'Times have changed. Perhaps these up-to-date activities have some educational value after all?' The wise old men were appalled: 'The essence of true education is timelessness. You must know that there are some eternal verities, and the sabre-tooth curriculum is one of them!'

I HAVE A VISION OF THE FUTURE, CHUM!

Successors to New-Fist in our own society, those who indulge as he did in the 'socially disapproved activity of thinking', will clearly recognize elements of a current debate in education. Chapman spoke for many of us when he said:

It is manifestly clear that what we have today is incapable of adaption to the future. There is no match between a curriculum structure

which, in essence, predates the invention of the printing press, relying
. . . on the teacher handing down knowledge from generation to
generation by word of mouth, and a society for whom access to, and
the processing of, information will require little more than the touch
of a button.[2]

Today's curriculum will not fit children for (or into?) life in the society of
the future—that seems to be the message on which we may all agree.
The remedy appears to be quite straightforward: a curriculum for the
technological age, preparation for work in 'hitech' industry and induction
into the 'realities' of society in an age of information.

How 'real' is our reality? There has been no shortage of prophets of
the new age, from the new economic order of Bell's 'Information Society'[3]
to the gushing 'gizmology' of Berry's 'Super Intelligent Machine'.[4] Perceptions of present economic decline have been offset by visions of a bright
computer-based future. The horrifying spectre of mass unemployment,
and the possibility of consequent social collapse, stand as reminders of
the stark alternatives to the new order. However, the pressure of the
situation appears to be such that short-term expediency is rationalized as
part of long-term social reconstruction.

There is a lack of true vision, both of the future and, perhaps more
importantly, of the present. This lack of perception is clouding our judgement. Our potential for exploiting technology runs a long way beyond
the current limits of technological development. We are mesmerized by
technology, like rabbits before stoats. Yet we command the technology.
Debate which considers only the impact of the computer on society misses
the point, and places the initiative on the wrong side. There is a two-way
interaction, as O'Shea and Self[5] observed: 'The effect of technology on
society is often discussed . . . but the effect of society on technology is
harder to forsee.'

The indications are that we can drive technology to meet strongly felt
needs—to put a man on the moon, or to wipe man off the face of the
earth—if that is what we really want. Our horizons should not be limited
by what is currently possible. We should be thinking now about what
education will need to be like in a high technology society. Yet everywhere
the signs are that our thinking on the nature of the computer and its place
in education are ossifying. Worse still, they appear to be ossifying round
visions of the future!

There is a certain point of view which tends to see the future as little
more than 'the present but more of it'. This is summarized in the admass
cliche which represents future societies in terms of shaven-headed women
dressed in white and surrounded by rectangular artefacts. This vision has
been remarkably stable for at least a generation, and only seems to vary
in terms of the technological 'frills' in the white-painted sterile 'blocks-
world' which these female paragons inhabit. I believe that many workers
in AI (or at least in fields where AI is a usefully prestigious buzzword!)

think of the future in terms of similarly uncluttered cliches, where only the technology in which they are personally interested is detailed, and where any attention to social context is minimal (and often naive). It is one of the strengths of SEAKE that it chooses to question this myopia.

If we as scientists and teachers are to influence the future, we must be aware that our research does not occur in a social and moral vacuum. The Auschwitz ovens were technological triumphs, but that is no credit to their designers. Nor can we claim credit if we allow ourselves to be diverted from questioning the social implications of what we are doing. The excitement (and kudos) of our activities in AI may distract our attention from the realities of the social models which our attempts at futurology may be used to validate. We may be unconsciously intervening on the wrong side of a long drawn-out ideological debate.

THE ELECTRONIC SABRE-TOOTH

The vision of the future which is most disturbing in its potential for ossification involves the response of education to the challenge of the microelectronics revolution. For a start, nobody is entirely sure what the challenge really is! The development of education for the 'information era' seems to be motivated by short-term political considerations, a general gut-reaction that something is wrong and a whole nexus of special interests. Whatever else, the future *must* include our bit of the present!

In particular, a link has been forged between the educational use of computers and computer-based information technology (IT) and some sort of economic utilitarianism. In Britain, when introducing the microelectronics education programme (since transmogified into microelectronics *in* education programme, MEP), the Minister responsible told Parliament that its function was '. . . to equip young people with the skills required to exploit the economic potential of the new technology'.

In fact, association between educational practice and economic progress has a long history. The bill, for example, which became the 1870 Education Act, enabling the public provision of mass education, was presented along with the sentiment that: 'Uneducated labourers are for the most part unskilled labourers, and if we leave our work-folk any longer unskilled . . . they will become overmatched in the competition of the world.'

This view, uttered in response to a nineteenth century perception of relative national decline, could be almost exactly mirrored in many contemporary discussions of the true significance of IT in education. It is perhaps worth considering, too, that whatever the 'humidity' of the government in power, and however firm its declared attachment to human values, it is usually those societal needs which are most directly economic in character which carry most weight in the counsels of the great and good. The concept of educational expenditure as 'infrastructural investment in human capital formation' (or some more humanitarian-sounding formulation of the same idea) exerts a powerful influence on the political mind of all persuasions.

People are people—not industrial raw materials—a point which may be missed amid the overtones of instrumentalism and economic ultilitarianism of current debate. Societies do no necessarily benefit if education is dedicated to short-term economic ends. Political rhetoric seems to demand that we admire and even emulate nations which are characterized by deeply entrenched social injustices, so long as they are economically successful. To our discredit, we do not seem to treat this idea with the contempt it surely deserves. Yet education should not merely be a means of 'regulating the quality of labour' or, at the individual level, merely a route of access to superficial competence in the great world outside. It is also about access to the inner world of the human mind.

It is not for any external agency to dictate where this 'access to the mind' is to lead. We cannot restrict education (as was tried in the nineteenth century) simply because we are fearful of the uses to which it will be put. Nor should we be foolish enough to erect barriers between 'practical' and 'theoretical' education (or training in whatever one chooses to call it). The very nature of work is changing. For the first time in history, society must maintain itself through the mass deployment of intelligence, rather than by the direction of the physical labour of the many by a small privileged minority of thinkers. The consequences for education are both radical and dramatic. Much of our rhetoric about the need for training fails to take this into account, though it was recently recognized by Burnett:[6] 'There is a current emphasis in Western education in favour of training over education. Yet training in the use of technology seems to reveal that we also require a strong humanistic perspective. The difficult questions are not 'how to. . . ?' but 'whether to. . . ?.' He went on to ask the very pertinent question of where we want our culture to go in the future, and whether what we are doing today is consistent with our visions.

There is a vital question of principle embedded here. It relates to the control of the use of technology, rather than simply to the control of technology. It also relates not to short-term expediency, but to the deep issue of the needs and rights of free citizens in a future society. The real questions to be faced in educational debate are not how to mechanize 'instruction', nor how to develop 'technological literacy', but how to help people develop themselves to the fullest assertion of citizenship in the face of an uncertain future.

If we concentrate on attempting simply to teach about technology and if we bend the curriculum to the apparent needs of the present (based on an analysis of the past and rationalized as anticipation of the future), we can expect to achieve nothing. Any curriculum which does not address itself to fundamental issues will fail—and the price of failure could be social catastrophe. Yet many of the curricular nostrums which are currently being peddled—by the AI community just as much as anybody else—have little relevance to the world of today, much less the world of the day after tomorrow.

For example, the current widespread emphasis on the study of computational devices seems fundamentally misguided, whatever the pressures may be which have made 'computer studies' one of the most popular school examination subjects in England. If the purpose of this subject is a sort of recent antiquarianism, it may at least have an elegant uselessness about it which may fit into the educational mosaic; but if its purpose is to teach about computing in any utilitarian sense, then it must be fundamentally misguided. Very few who teach it have any depth of grounding in its substantive discipline (in the United Kingdom even university faculty are difficult to recruit because of low salaries compared with the business world). What is more, its taught content will inevitably be obsolete or obsolescent. Aleksander and Burnett[7] made this point: 'Teaching people to make current computer structures and to program them, when the research community is endeavouring to alter such structures out of recognition and to replace programming by more natural means of communication . . . seems sheer lunacy.'

However, that is just what we are doing. We are building on electronic sabre-tooth curriculum which is obsolete before it is even implemented. We are also associating it with an entrenched bureaucracy and with the personal interest of all the individuals who have contributed to it.

'NEW-FIST, THOU SHOULDS'T BE LIVING . . .'

Although the term 'sabre-tooth curriculum' has become a byword for curricular obsolescence, it should not be forgotten that New-Fist's educational ideas were initially brilliantly successful. It was the system which failed, rather than the curriculum. The idea of relating educational processes to desired objectives (whether intellectual, social or economic) is as valid now as it was in the paleolithic era. However, our world is closer to the world of New-Fist's innovative successors: the institution of education has come into being, and its very existence alters the social environment. Moreover, the institution has become associated with all sorts of objectives, few of which are truly related to the overt business of education.

When we contribute ideas, devices, media, materials or whatever to the present educational structure, whatever *our* motivations or inspiration, we are contributing to the continuance of something which is, for the vast majority of people, an agent of repression, rather than liberation. This was put very clearly by Edwards[8] in his inaugural address at Newcastel University. He pointed to the role of education as an agent of what he described (after Durkheim) as the 'ideological state apparatus'. Teachers are unconsciously or otherwise part of this apparatus—as are AI researchers. This apparatus is more closely linked with the needs of the state, rather than of individual citizens, and, as in any such system, the needs of the state are operationally defined in terms of the power needs of those who most benefit from its operation. This has always implied some attention to manpower needs, often at the expense of education:

> It is manifest that in a free nation, where slaves are not allowed, the surest wealth consists in the multitude of laborious poor. . . . To make society happy and people easy under the meanest circumstances, it is requisite that great numbers of them should be ignorant as well as poor. . . . A man who has had some education may follow husbandry by choice . . . but he won't make a good hireling . . . for a pitiful reward (Bernard Mandeville, 1660–1736).

This is a sentiment not far removed from the current political arena!

Edwards argued that liberal-egalitarian hopes are bound to fail where the nature of the education offered and the dominant definitions of ability and achievement systematically favour those with the necessary cultural inheritance. I shall go further and suggest that any attempt at all to develop education to serve human needs must fail where the whole essence of the system is dedicated to serving the utilitarian objectives of a small subset of the population. Most 'reforms' of recent years may be seen as attempts to enlist the acquiescence of the many to the service of the sociocultural objectives of the few.

Education in Britain, whatever its rhetorical basis, appears to be directed towards the conservation of social differentiation through a variety of mechanisms, including the regulation of expectation and control over the flow of knowledge and information. The privileged position of the universities at the apex of a hierarchically structured system (reflecting the structure of society at large) has allowed them to impose, through the vicarious action of a subtly regressive examination system, a concept of knowledge which is every bit as ossified as that enshrined in the failing sabre-tooth curriculum. We have a system of education attuned to the knowledge requirements of a highly stratified society and the manpower needs of the feudal system, justified in terms of the transmission of culture and/or the maintenance of 'standards'. It is into this environment that we are now introducing modern IT. If we simply use this as a technological 'fix', then we will fail. It is no longer a matter of fine-tuning an essentially satisfactory system, but of realization that the whole system is beyond economic repair!

It is not enough for today's New-Fists to look for new utilitarian objectives to replace the old. This has been the greatest failing of vision among researchers in AI, as elsewhere. New delivery systems for education, new media, new methods are irrelevant if the basic social environment is left unchallenged. For education, by its very presence as a system, changes the structure of society in ways that it would be naive and impractical to ignore. Education as an enterprise (consuming, after all, some 6 per cent of GNP annually) is a massive part of society, and it is the whole structure which needs reform, not its peripheral operations—as Hubbard[9] realized, the dilemma we face is between the pursuit of patterns of the past or a recognition that new times require new means.

Let me conclude by repeating my earlier comment. What we need is New-Fist's original clarity of insight and innovative thought, not a series of look-alike adaptations to notional and probably non-existent situations. It is no use arguing for closer links between education and industry, '. . . far more must be done to link our schools with industry, starting not at 14 but at 11 or 12 years. . .'.[10]

We have trodden that path before, and it has failed because it has refused to tackle the basic assumptions on which the system is founded. Nor for the same reasons is it any use our providing even more elaborate 'technological pit-props'. We cannot deny the reality of the economic imperative, but the last word must lie, ironically, with the tribal elders who upheld the irrelevant sabre-tooth curriculum: '. . . the essence of true education is timelessness . . . there are some eternal verities. . . .' Even if they were wrong in identifying their curriculum with these verities, they were surely right in asserting values outside of immediate utilitarianism.

We in the AI community are in a strong position to contribute to the *new* education. We have much to overcome:

We have no use for theological subtleties
The beliefs we have inherited, as old as time
Cannot be overthrown by any argument
Nor by the most inventive ingenuity.[11]

But we can try!

REFERENCES

1. Benjamin, H. (1975). 'The saber-tooth curriculum'. In *Curriculum Design* (Eds M. Golby, J. Greenwald and R. West), Open University Press, Milton Keynes, pp. 7–14.
2. Chapman, B. (1983). 'Problems, perspectives, paradoxes, possibilities'. In *Curriculum Comment* (Ed. J. Craig and T. Adams), MEP, Newcastle-upon-Tyne.
3. Bell, D. (1980). 'The social framework of the information society'. In *The Microelectronics Revolution* (Ed. T. Forester), Blackwell, Oxford, pp. 500–549.
4. Berry, A. (1983). *The Super Intelligent Machine*, Jonathon Cape, London.
5. O'Shea, T., and Self, J. (1983). *Learning and Teaching with Computers*, Harvester Press, Brighton.
6. Burnett, D. (1984). 'Logo for teacher education'. In *New Horizons in Educational Computing* (Ed. M. Yazdani), Ellis Horwood, Chichester, pp. 72–83.
7. Aleksander, I., and Burnett, P. (1984). *Reinventing Man*, Penguin, London.
8. Edwards, A. D. (1980). 'Schooling, liberation and repression. Unpublished inaugural lecture, Newcastle University.
9. Hubbard, G. (1984). 'Social and educational effects of technological change'. *British Journal of Educational Studies*, **32**(2), 108–117.

10. British Computer Society (1984). *Curriculum for the Future*, BCS, London.
11. Euripides (1954). *The Bacchae* (translated by P. Vellacott), Penguin, London.

Artificial Intelligence for Society
Edited by K. S. Gill
© 1986 John Wiley & Sons Ltd

18.

ARTIFICIAL INTELLIGENCE AND EDUCATION: A CRITICAL OVERVIEW

MASOUD YAZDANI Department of Computer Science, University of Exeter

ABSTRACT

In this paper we present an overview of the contribution of artificial intelligence to education. Major areas of application are identified and described, including intelligent tutoring systems, AI programming environments and microworlds. We also point out the use of PROLOG as a method of using databases in education, as well as the use of expert systems for tutoring. We single out the work in user modelling and machine learning and courseware design to be the areas where more work is required in order for these systems to be educationally more effective.

INTRODUCTION

One of the major areas of human endeavour is education. It is therefore not surprising that from the early days computers have been put to use in education. Very rapidly their use has been in two distinct areas: on the one hand, computer assisted learning (CAL) packages help teachers to 'teach' different subjects in the school curriculum and, on the other, the writing of computer programs (as this happens mostly in BASIC) teaches children now a computer works.

Accepting that computers are the latest tools invented by man, we can see that they are being used in the same way as many other such tools. Children during programming learn how to use these new tools, while teachers use them in the same way as they use the blackboard: to simplify their teaching task. There are good historical reasons why educational computing, up to now, has been dominated by BASIC and CAL. Computers have been rare and therefore they could only be used by the privileged class in any environment (e.g. teachers). They have also been expensive, so the computers which schools could afford would only support programming languages (e.g. BASIC designed to optimize the machines' efficiency, instead of providing for the needs of human users.

In the meantime, a new force has been active in bringing about some changes to the scene. Artificial intelligence (AI) has concerned itself with the study of giving computers abilities usually associated with human beings (such as the ability to understand natural language, solve problems, play games), as well as to learn for itself. The results of more than 25 years' work in AI are now beginning to affect many human endeavours, including education.

In this paper we look at a number of ways that both the techniques and ideas from AI have influenced educational applications of computers.

INTELLIGENT TUTORING SYSTEMS

Intelligent tutoring systems[1] are AI's answer to CAL packages. While CAL has tended to be basically drill and practice, intelligent tutoring systems (ITS) have aimed to be diagnostic. The following incorrect subtraction and addition

$$
\begin{array}{r}
170 \\
-\ 93 \\
\hline
187
\end{array}
\qquad
\begin{array}{r}
33 \\
+179 \\
\hline
102
\end{array}
$$

will not result in the message 'wrong, you lose a point' being printed on the screen, but will lead to a correct diagnosis of the pupil's error in forgetting the horrow or the carryover.

These systems succeed by containing clear articulation of knowledge involved in a narrow domain. One such system, DEBUGGY,[2] performs as well as, or rather better than, human teachers in diagnosis of misconceptions of pupils when performing subtraction.

There have been two major criticisms levelled against such sophisticated systems. Such systems 'have not yet been incorporated within a remedial program, with which students can interact to improve their subtraction skill; nor has it yet been presented in such a form as to be usable as a diagnostic aid by any mathematics teacher'.[3] The major reason behind these shortcomings has been the complexity of the task involved. Even in such a narrow domain, such as subtraction, there are numerous ways in which a pupil can make mistakes. Therefore a program such as DEBUGGY would be beyond the power of a modest school microcomputer. However, as the cost of hardware is declining, it has become possible to offer some of this level of sophistication to the school teacher. Attisha and Yazdani have produced a system which uses a taxonomy of possible errors which children make in addition and subtraction in order to provide remedial advice, similar to that of DEBUGGY, using a school microcomputer.[4] Furthermore, Attisha and Yazdani have extended this work to cover multiplication, which is by nature more complex than subtraction.[5] In multiplication the pupil could make mistakes due to various reasons: problems with the multiplication table, with the multiplication algorithm or with the addition of subtotals. When errors in any two of these areas are combined, the result could appear to be nothing more than carelessness (random) to the best of human teachers. However, the computer system provides exercises in order to isolate different areas of difficulty and diagnose the problem.

ARTIFICIAL INTELLIGENCE PROGRAMMING ENVIRONMENTS

Intelligent tutoring systems are rather effective for the teaching of narrow domains but construction of systems based on general competence in areas such as 'language development' or 'problem solving' seem impractical. Instead, Papert has argued in favour of the development of 'learning environments' which provide a student with powerful computing tools.[6] The student thereby engages in an open-ended learning-by-discovery process by programming the computer to carry out interesting tasks, although it is argued that the intention is not to learn how to program a computer but to learn through programming a computer.

Artificial intelligence programming environments are tailored to human beings, minimizing the cognitive load put on a naive user, as opposed to optimizing the machine's efficiency. Yazdani presents four such environments based around LOGO, SOLO, PROLOG and POP-11.[7] The practical contribution of such environments is that they make it possible to design computing systems which are educational, fun and which relate to childrens' basic feelings.

Some advocates of the use of AI programming environments would go even further than that. They argue that by encouraging as many people as possible to write programs as AI scientists do, we would 'introduce them to ideas about themselves, their minds, and the universe in which they find themselves which they might otherwise not have encountered'.[8]

Pupils can attempt to use AI programming environments in order to produce simple programs which have a certain level of 'intelligence'.

They move from the position of being experimented on to doing the experimentation themselves. By trying to write a program which learns to play a game of noughts and crosses, the pupil starts to realize what learning is.

MICROWORLDS

Papert has argued that such activity (playing with LOGO, for example) has a similar role to that of playing with sandpits in the Piagetian theories of learning.[6] In playing with AI programming environments children build 'objects to think with' in place of sand castles. This building process takes place in what is known as a 'microworld': a limited portion of the real world whose characteristics can be easily understood.

AI is fond of using microworlds in all its areas of research: microworlds can easily formalized. The most influencial AI work in natural language processing[9] converses with the users about a small world of a table top with a number of coloured boxes inhabiting it. The computer program is capable not only of obeying orders in this world but also of discussing it in detail. Another program[10] will learn new concepts from the old if taught by a human teacher. The obvious hope of researchers has been to link all these programs together and let the computer learn how to deal with the real world, in the same way as a child playing with blocks. Unfortunately AI has not yet been able to translate its successful attempts with microworlds to the real world. There seems to be such a level of increase in complexity of the domain when moving away from the microworlds to the real ones that most lessons need to be relearned.

Papert has argued that such microworlds would constitute a very good complement to the ones usually used by children up to now (such as Meccano sets) and would be as effective, if not more.[6] The most well known of these computational microworlds has been turtle graphics. This is a computing package (based around a mechanical device with the same name) which will take commands from children in a way similar to their pet 'turtle' if it could understand them. In this way children succeed in drawing wonderful shapes by giving simple commands to the computer. In addition to this a number of other microworlds have been successfully used, such as one based around a beach,[11] one based around the world of a farmer trying to transport a fox, a chicken and a bag of grain across a river.[12] The possibilities in this area seem to be endless and already a large number of practising teachers have taken up the challenge.

PROLOG AND DATABASES

Building and querying simple databases of interesting information, when combined with PROLOG's capability to draw higher level inferences from the facts it has been given, seems to offer a very strong educational tool.[13] The pupil not only learns the information contained in the database with enthusiasm but also learns logical thinking.

PROLOG, for example, has been used to set up a database of information about particular historical events, and children then query the database in order to discover the cause of the events. Nichol and Dean argue that this would give the children the 'feel' for what it is to be a historian looking for information, sifting the evidence.[14] In addition to acting as a database of information, with features missing from fourth generation commercial packages, PROLOG can also be viewed as a general-purpose AI programming language. The difference is, however, in the style of programming. Something which Kowalski has argued[15] in itself is an important turning point. In languages such as BASIC, or even LOGO, the user has to explain to the computer 'how' to perform actions but when using PROLOG the user only needs to specify 'what' is to be done. The decision how to do the task is left to PROLOG's own powerful search mechanism.

EXPERT SYSTEMS

AI's contribution to education does not end at enriching the previously existing components. It also introduces new ways of learning and teaching. One novel way of teaching a subject such as physics or medicine would be to produce an expert system which would behave like a skilled physicist or a medical consultant.[16] It is claimed that the trainee can then observe the knowledge and the line of reasoning of the program and learn by it so that trainee doctors, for example, could simply be asked to look over MYCIN's[17] shoulder as it sets about solving its problems. This is because:

1. MYCIN can explain in English what it is doing.
2. MYCIN's decision-making processes are similar to those which students are supposed to develop.
3. MYCIN's representation of medical information is in a human-like manner.

MYCIN is one of a number of AI's commercially viable propositions called an expert system. Such systems are experts in a very narrow domain of knowledge to a degree that, within their domain, they can match the performance of human experts, and possibly exceed them.

The GUIDON program[18] is one application using existing expert systems such as MYCIN for educational purposes (see Fig. 1). Clancey argues that although MYCIN-like rule-bases are a good means of transferring knowledge from a human expert to a human trainee; the rules in themselves are not sufficient. He proposes adding two further levels: one a 'support' level to justify rules and the second an 'abstraction' level to organize rules into patterns in order to transfer such systems into a tutorial medium. Further, he argues that such systems still need teaching expertise of a general kind and natural language competence to carry out a coherent dialogue with the student.

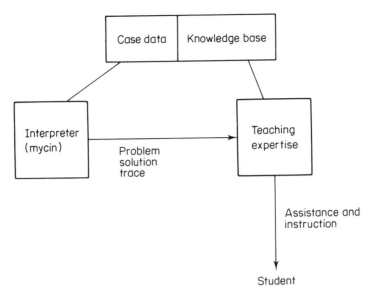

Fig. 1   The overall structure of GUIDON

Clancey's work is encouraging in the task of providing computer-assisted instruction in a domain where the system itself is capable of performing the task that it is expecting the trainee to perform. For example, we have chosen a new domain of application (language teaching) which, to our knowledge, has not yet been the subject of such treatment.[19]

Most language teaching programs rely on a massive store of correct sentences and divisions from them. Despite the large number of legitimate constructs with which they can deal, the programs are really nothing more than a dumb, if effective, pattern matcher, linking unintelligible orders of characters to those pre-stored. As a consequence, the programs cannot recognize or comment upon errors encountered, even if the errors are frequent, unless these have been individually and specifically anticipated by the programmer. Therefore, the standard of accuracy required (coupled with the time for preparation of exercises) seems very high indeed.

Instead, systems such as FROG[20] and FNLGA,[19] which use a general-purpose language parser, can cope with an indefinite number of possibilities without being programmed in advance to anticipate all the possibilities.

IMPORTANT UNSOLVED PROBLEMS

So far in this paper we have identified five major distinct ways where artificial intelligence ideas (such as microworlds) or AI techniques (such as expert systems) can be taken to advance the more standard applications of computers in computer assisted learning and computer literacy. However, it should be noted that while these are all very promising, none of them is as yet free of unsolved problems. What follows is my personal view of the major unsolved problems.

Machine Learning

Most intelligent tutoring systems rely on a good knowledge of the domain where the teaching takes place. For example, in our own system for subtraction,[4] we relied on a vast amount of existing data in order to build a full taxonomy of children's errors in subtraction. The problem in fact is also suffered by tutoring systems which use production rules or procedural networks which, although more powerful, still rely on a reasonable knowledge of the domain while they are being constructed. Such knowledge is not readily available in cases where we have a more complex domain than arithmetic.

What seems to be obvious (but rather hard yet to contemplate using) is that of 'machine learning'. The work in this area is already reasonably advanced[21] in order to incorporate techniques which would enable a computer tutor, similarly to a human tutor, to learn more about the domain in which it is teaching. However, such sophistication requires computational resources well beyond those available to educational establishments.

User Modelling

Instant feedback and individualization have been the twin gods of computer-assisted instruction[16] for three decades. AI systems seem to provide the possibility of not only instant feedback but also a reasonably rich level of it, good remedial advice for example. However, on the individualization side, AI does not yet seem to have offered any advances on what CAL had done. GUIDON, for example, suffers from the fact that it is always comparing a trainee doctor's actions with those of MYCIN while what is needed is a knowledge of what a particular user knows about the topic, his or her current misconceptions in addition to the terminology with which he or she is familiar and the forms of explanation which are found to be effective.

These needs are shared by most interactive AI systems and a good deal of research is already in progress in what is labelled 'user modelling'. My personal view is that these problems can be viewed as extensions to the 'machine learning' ones. Human teachers are constantly learning, both about the domain in which they are teaching as well as the pupils they are teaching.

COURSEWARE DESIGN

Using sophisticated AI programming languages, as well as their supplementation with educational microworlds, provides an exciting possibility of using them for learning abstract concepts such as problem solving.

However, concrete evidence has yet to be given that this process is in fact educationally effective. What is an open question is whether the pupil who learns powerful new ideas while playing with a microworld could, in fact, reapply them in the real world. Lighthill pinpoints the obsession with microworlds in his most effective critique of AI:

> Most robots are designed from the outset to operate in a world as like as possible to the conventional child's world as seen by man: they play games, they do puzzles, they build towers of bricks, they recognise pictures in drawing-books ('bear on rug with ball'); although the rich emotional character of the child's world is totally absent.[22]

The major weakness of microworlds and the free-learning environment idea on which they are based, is exactly what makes them so powerful. They can cope with a large variety of possibilities, limited by the imagination of the user. The user is free to chose his or her own problems and the route to their solutions. This approach, whilst increasing the pupil's motivation, simultaneously decreases learning efficiency. The pupil is simply not very good at selecting his or her own learning strategy. However, this shortcoming can be alleviated by designing 'courseware' within which such activity is fitted. Here a human teacher holds the pupil's hand when necessary and leaves him or her free to explore when educationally effective. In other words, the shortcoming has so far been in viewing these systems as complete environments. However, they are in fact elements of a larger human environment where the emotional context is provided by the human teacher.

CONCLUDING REMARKS

The advocates of the use of AI in education seem to be grouped in two camps representing the two ends of a spectrum into which we could place the areas mentioned in this paper. One group concerns itself with more powerful 'teaching packages',[1] which teach a subject aiming to be as competent and as structured as a professional human teacher.

At the other end of the spectrum, typified by,[6] researchers are working on 'learning environments' which provide students with powerful computing tools with which they can learn through play. The arguments presented in the previous section point to the fact that neither of these are yet capable of being used effectively in education. What seems to be missing is a happy medium in this spectrum. I believe it is possible to build systems which can incorporate good features of both approaches. For example, to teach grammar of modern languages, we could use specific tutoring systems (such as that of Barchan, Woodmansee and Yazdani[19]) for beginners. We could also provide more advanced students with the tools which we used to build the tutor. Advanced students could then build tutors for teaching beginners. In this way the advanced students learn by trying to be teachers. As they say in educational establishments, 'the best way to learn a subject is to teach it'.

I believe that with further advancements in AI theory and techniques we shall see that the distinction between the different approaches mentioned in this paper are more conceptual than real. In the long term these different components will merge by incorporating some or all of them, depending on the subject matter and computational power available to educational establishments.

REFERENCES

1. Sleeman, D., and Brown, J. S. (Eds) (1982). *Intelligent Tutoring Systems*, Academic Press.
2. Burton, R. R. (1982). 'DEBUGGY: diagnosis of errors in basic mathematical skills'. In *Intelligent Tutoring Systems* (Eds D. Sleeman and J. S. Brown), Academic Press.
3. Boden, M. A. (1982). In *The Educational Implications of Artificial Intelligence* in (Ed. W. Maxwell), Thinking: the New Frontier Franklin Institute Press, Pittsburgh.
4. Attisha, M., and Yazdani, M. (1983). 'A microcomputer-based tutor for teaching arithmetic skills'. *Instructional Science*, **12**.
5. Attisha, M., and Yazdani, M. (1984). 'An expert system for diagnosing children's multiplication errors'. *Instructional Science*, **13**.
6. Papert, S. (1980). *Mindstorms—Children, Computers, and Powerful Ideas*, The Harvester Press/Basic Books.
7. Yazdani M. (Ed.) (1984). *New Horizons in Educational Computing*, Ellis Horwood Ltd/John Wiley and Sons.
8. Coker, M. (1984). 'Creating a 'good' programming environment for beginners'. In *New Horizons in Educational Computing* (Ed. M. Yazdani), Ellis Horwood Ltd/John Wiley and Sons.
9. Winograd, T. (1972). *Understanding Natural Language*, Edinburgh University Press.
10. Winston, P. H. (1975). 'Learning structural descriptions from examples', in Winston, P. H. (ed) *The Psychology of Computer Vision*, McGraw-Hill, New York.
11. Lawler, R. (1984). 'Designing computer-based microworlds'. In *New Horizons in Educational Computing* (Ed. M. Yazdani), Ellis Horwood Ltd/John Wiley and Sons.
12. Sloman, A. (1984). 'Beginners need powerful systems'. In *New Horizons in Educational Computing* (Ed. M. Yazdani), Ellis Horwood Ltd/John Wiley and Sons.
13. Ennals, R. (1984). 'Teaching logic as a computer language in schools'. In *New Horizons in Educational Computing* (Ed. M. Yazdani), Ellis Horwood Ltd/John Wiley and Sons.
14. Nichol, J., and Dean, J. (1984). 'Pupils, computers and history teaching'. In *New Horizons in Educational Computing* (Ed. M. Yazdani), Ellis Horwood Ltd/John Wiley and Sons.
15. Kowalski, R. (1984). 'Logic as a computer language for children'. In *New Horizons in Educational Computing* (Ed. M. Yazdani), Ellis Horwood Ltd/John Wiley and Sons.
16. O'Shea, T., and Self, J. (1983). *Learning and Teaching with Computers*. The Harvester Press.

17. Shortcliffe, E. H. (1976). *Computer-Based Medical Consultations: MYCIN*, American Elsevier.
18. Clancey, W. J. (1979). 'Tutoring rules for guiding a case method dialogue'. *International Journal of Man–Machine Studies*, 11.
19. Barchan, J., Woodmansee, B., and Yazdani, M. (1985). 'Computer assisted instruction using a french grammar analyser'. Research Report No. R. 126, Department of Computer Science, University of Exeter.
20. Imlah, W. G., and du Boulay, J. B. H. (1985) FROG, Department memo, Congnitive Studies Programme, University of Sussex.
21. Michalski, R. S., Carbonell, J. G., and Mitchel, T. M. (1984). *Machine Learning: An Artificial Intelligence Approach*, Kaufmann Inc./Springer Verlag.
22. Lighthill, J. *et al.* (1973) *Artificial Intelligence: A Paper Symposium.* Science Research Council, London.

Artificial Intelligence for Society
Edited by K. S. Gill
© 1986 John Wiley & Sons Ltd

19.

LOGIC PROGRAMMING AND EXPERT SYSTEMS

CHRIS MELLISH Cognitive Studies Programme, University of Sussex, Brighton

ABSTRACT

Constructing an expert system involves creating a computer system that has human-like expertise and exhibits intelligent behaviour in some restricted domain. Such a system often has comprehensive knowledge about its application domains, and this knowledge often has to be encoded and refined in a long (perhaps semi-automatic) development phase. Current programming languages and environments do not always facilitate this. Many people now argue that for building expert systems we need to depart from traditional data processing models of computation and seek new architectures. The concept of *rule-based systems* has been often promoted in this context. Rule-based systems offer the possibility of encoding knowledge in a clear modular fashion which is nevertheless executable. However, many rule-based systems do not have a clear semantics and rely on side-effects which destroy modularity. *Logic programming* can be viewed as a kind of rule-based programming which seeks to remedy these defects. In logic programming, the semantics of rules it well understood and the modularity of rules guaranteed. This is because the rules are written in logic.

FROM DATA PROCESSING TO RULE-BASED SYSTEMS

According to a certain traditional view, a computer program is a set of instructions, expressed in a special programming language. The machine uses these to operate on a set of data, specifying a particular problem at hand, to produce some kind of result. Let us look at an imaginary 'tax advice' program in this light. Each time the program is run, it is presented with data specifying a particular person's financial circumstances, and it prints out suggestions about how that person could save on their tax bill. Among other things, the program might indicate whether it would be better to have husband and wife taxed separately or together, and whether the person could gain by having certain payments made by deed of covenant. Thus, one part of the program might specify a computation to determine the tax due with husband and wife taxed together, another part might determine whether the person is liable to super tax, and so on. Such a program would be quite easy to construct and obviously useful. However, what happens when the tax rules change? The trouble is that a fairly simple change (such as a change in the tax thresholds) might well affect quite disparate parts of the program—it might affect both the calculations of tax involving a married couple and also the calculations involving deeds of covenant, for instance. In general, it will be quite hard to track down the parts of the program that depend, directly or indirectly, on a given change in the regulations. Of course, a programmer can try to minimize this problem by writing a clear, well-structured program in the first place, but it would be very hard to envisage in advance all the kinds of changes that might arise, and most programmers would be unable or unwilling to do this. Another problem with the 'tax advice' program as formulated is that it is written in a conventional programming language, where *what* the program knows is all mixed up with *how* the program is to perform certain tasks. After all, the program is a set of instructions, not a textbook. Thus it would be hard for a non-programmer (or even many programmers!) to inspect the program code and determine whether this version of the program is actually taking account of the latest changes in tax law or what the program knows about life insurance policies. The program's knowledge cannot be easily accessed or assessed, as it is bound up in a *procedural* representation.

In a rule-based system, what was previously thought of as the 'program' is now seen as having two components, the *knowledge base* and the *inference engine*. Meanwhile, the input data to the system is still seen much as before (it is often called the *database*). The knowledge base is a data structure consisting of a set of *rules* expressing in small, independent chunks the system's knowledge. The inference engine is an actual set of instructions (a conventional program) to enable the machine to manipulate the rules and hance solve problems. Here is an example, in an English-like syntax (of course, actual rule-based systems used formalized syntax), of a rule that might be in a system to play 'noughts and crosses' ('tic-tac-toe'):

> If
>> there is an 'X' in square A
>> AND there is an 'X' in square B
>> AND square C is empty
>> AND square A-B-C form a line
> THEN
>> put a 'O' is square C

This rule expresses a single piece of strategy/knowledge about 'noughts and crosses'. In general, one would expect a rule-based system that played the game well to have a number of such rules, specifying possible actions to take in different situations. It is up to the inference engine to determine which rule to use when in a given game. Usually it will not matter much how the inference engine works (it might well be bought as a 'shell' written by someone else), as long as the actions it takes are faithful to the principles embodied in the rules. For instance, one would not expect any action to take place that did not result from the conditions of a rule being satisfied. The inference engine can be an impenetrable piece of program in the same way that a compiler for a high level language can. Most users of a compiler do not bother to question how it works, as long as it respects the semantics of the programming language.

The claimed advantages of rule-based systems over conventional programs are perspicuity and modularity. These systems tend to be perspicuous because the knowledge can be directly examined. This is very important for system maintenance and debugging, and also leads to possibilities such as the automatic generation of explanations. Modularity results from the fact that the knowledge is split up into small, independent rules. Thus it is relatively easy to change a rule-based system by adding, altering or removing a single rule. Modularity also leads to the possibility of systems that can learn by themselves to some extent.

FROM RULE-BASED SYSTEMS TO LOGIC PROGRAMMING

Although it is possible to construct rule-based systems that have the claimed advantages of perspicuity and modularity, in practice systems developed often do not have these advantages. This is partly because there are a number of features in rule-based systems that compromise the independence of rules. The main problem is that rules of the form given above are about actions and causing changes in the world. Since the applicability of a rule depends on the state of the world, in general the set of rules that are applicable at some point depends on the rules that have run previously. If the rule writer is not careful, when the system runs one rule it may cause a change which prevents another rule running which would otherwise have been applicable. In such a situation, the order in which the inference engine decides to run the rules becomes very significant. One common response to this sort of problem is for the programmer to add extra actions and conditions to rules which ensure that particular rules run in a particular order. This is, however, contrary to the spirit of rule-based programming and returning to the conventional motion of providing a sequence of instructions.

If a rule-based system maintains a database consisting of the 'facts' it knows and the actions performed by its rules are restricted to 'concluding' new facts, these sorts of problems will not arise. In such a system, an applicable rule will always be applicable, and cannot be prevented from running by an action from another rule. Such a system is essentially a *pure inference* system. That is, the job of the inference engine is to work out what follows from the initial facts it is given, using the rules as rules of inference. This is one of the basic ideas behind logic programming.

A logic programming system can be regarded as a kind of rule-based system where the database and the rules are logical axioms (in fact, the only criterion for separating these two parts is that the rules express the system's general knowledge and the database expresses the particular knowledge appropriate to the problem at hand). In such a system, the inference engine is a *theorem prover*, i.e. a machine for discovering logical statements that follow from the given axioms. The advantage of using logic as the language for our rules and database is that in logic the notion of 'following from' is well understood and precise.

Unfortunately, the state of the art in automatic theorem proving has not produced theorem provers that are able, given arbitrary logical axioms, to discover all the 'interesting' things that follow without getting bogged down in irrelevances. Indeed, there are theoretical reasons to assume that this will never be possible. So, for a practical logic programming system, it is necessary to give the theorem prover hints, *control information*, in addition to the rules and database. These hints might involve suggesting that certain rules be tried before others, certain possibilities not be explored, and so on. Thus we have Kowalski's well-known 'formula':

Algorithm = logic + control

Whereas, in a general rule-based system, imposing different control regimes may result in different (even contradictory) answers, the idea in a logic programming system is that correct logical axioms will guarantee that any results produced are correct. The only effect that the control component can have is to change the order in which solutions are found, change the difficulty in finding a solution or (possibly) affect whether a given solution is found at all.

Here is a version of that 'noughts and crosses' rule, expressed in the way a logic programming system might see it (again, I have shown it in an informal, rather than a formal, syntax):

IF
> square A has a 'X' in board position S
> AND square B has a 'X' in board position S
> AND square C is empty in board position S
> AND squares A-B-C form a straight line

THEN
> a best move from position S results in a position S1
> AND S1 is identical to S,
>> except that square C has a 'O' in position S1

Notice how this differs from the previous version. Instead of talking about conditions and actions, this rule talks about what follows from what. In order to be a precise logical statement, the rule has to make explicit things that were only implicit in the previous rule (the fact that all the conditions have to be true of the same board position, for instance). In terms of functionality, however, this rule could do the same job as the previous one. To see what to do next in a 'noughts and crosses' game, we can execute the action suggested by the rule (in the ordinary rule-based system) or we can investigate which board states S1 are the result of a best move from the current position (in a logic programming system).

In a way, the advantages of a logic programming system come from the discipline it imposes on the rule writer. Being expressed in logic, the individual rules must be precise and independent—no rule can reference anything outside itself. A logic program is, in fact, closer to what is normally considered a *specification* of a task than a conventional program. The use of logic as the rule language means that there is a clear semantics and that the program is amenable to various treatments, such as correctness proving and automatic transformation.

WHERE DO WE GO FROM HERE?

The above discussion has suggested a rosy scene, with logic programming about to transform the way computer software is produced and regarded. The truth is, however, more complex, as there is not yet a programming system that lives up fully to the .ideals of logic programming. The most frequently used 'logic programming language' is Prolog (and its variants, such as MicroProlog), and Prolog is deficient in a number of ways when regarded as an embodiment of the logic programming aim. For instance, Prolog allows the programmer to create side-effects which destroy rule independence. It also provides facilities which can encourage the programmer to write inadequate logical statements that will be compensated for by appropriate control annotations. Thus we still have to regard logic programming as a long-term aim, towards which Prolog is only a first step. Progress has certainly been made, for it is possible to write elegant, pure, logic programs in Prolog and to have them executed efficiently. It remains to be seen how much further we can get towards a pure logic programming language without sacrificing efficiency excessively.

Another most important issue that we have not touched on is the potential accessibility of logic programming to non-programmers. One is tempted to conclude that logic programming is a higher level activity than conventional programming and that the logic programmer does not have to worry about implementation details in the same way as the conventional programmer. Therefore it should be relatively easy for a novice to produce and understand logic programs. Indeed, Kowalski has claimed that: 'Logic is . . . more human-oriented than other formalisms specifically developed for computers.'

However appealing this idea may be (and logic was indeed originally devised to explain human argumentation), we should be wary of assuming that people have any more inherent ability to precisely describe a problem than they have ability to precisely specify a set of instructions for how to solve it. A lot more research needs to be done before we can draw balanced conclusions about the human side of logic programming.

FURTHER READING

This article has made use of ideas from a number of people but has been a necessarily superficial account. The reader who wishes to learn more about logic programming and Prolog is referred to the following books:

Clocksin, W., and Mellish, C. (1981). *Programming in Prolog*, Springer-Verlag.
Ennals, R. (1983). *Beginning Micro-Prolog*, Ellis Horwood.
Hogger, C. (1984). *Introduction to Logic Programming*, Academic Press.
Kowalski, R. (1979). *Logic for Problem Solving*, North Holland.

Artificial Intelligence for Society
Edited by K. S. Gill
© 1986 John Wiley & Sons Ltd

20. REALITIES OF EDUCATIONAL SOFTWARE

BRYAN SPIELMAN Writer and Educational Software Publisher, Brighton

The use of computers in education is at last becoming customary. It is of course early days yet and there is a long way to go before the true immensity of the capabilities of computer methods becomes properly evident in any commonplace educational application. For the present, however, there is a great deal of value that can be achieved even within the limitations of equipment, expertise, time and money currently available to schools and homes. What has so far been realized falls a long way short of what is possible. Given extant resources, far from ideal though they are, it is still a case of 'could do better'.

My attention here is confined to educational software which runs on microcomputers and is intended to do its job in the presence of a learner.

First there are a number of common acronyms which it might be as well to get out of the way (and indeed get out of the way of). Among them may be mentioned the following:

1. CML, CAMOL mean computer managed learning, computer-aided management of learning. Futuristic or optimistic ideas here. Supposedly, a student's performance, achievement and progress through a course unit, or even a whole course, is tirelessly watched over, steered, assessed and recorded, liberating the teacher from clerical chores. Some

people find the prospect disturbing (though you would be more likely to get the hump with CML than with CAMOL).

2. CBL, CBT, CBT mean computer-based learning, computer-based teaching, computer-based training. The trouble with these is the word 'based'. It assigns to the computer a more central role than is healthy. Better to have HBL (human-based learning). T for teaching ought to be viewed with suspicion also—L for learning should suffice since if there is any teaching going on one would hope that some learning might be taking place consequentially. Training is an altogether narrower matter than education (cf. CAI below).

3. CAL, CAI mean computer-aided learning, computer-aided instruction. A for aided—or assisted—is greatly to be preferred to B or M (see above) since it puts the computer in its proper place along with other educational aids as something to aid or assist the process, much as the blackboard used to (BAL) or the overhead projector (OPAL), and casts it as just one further—but outstandingly enriching—resource. CAI is a subspecies of CAL and actually means programmed learning in the sense originated by B. Skinner.

From here on my observations will relate primarily to CAL. Moreover, it is only the form that I am considering, not the content, other than by way of illustration. The two cannot be entirely separated, of course, since the content of a piece of software often dictates the form while the form may place constraints on the content. Most of what I have to say applies to CAL software whatever its content.

CAL software can come in a variety of forms, irrespective of content, and what might be an appropriate attitude to take when considering one sort could be inappropriate when considering another. Critics, reviewers, advisers, experts and other well-wishers often pass judgements which are entirely misapplied. Before embarking on writing or using or evaluating a program it will save a great deal of dismay if you first establish the species of program you are concerning yourself with.

A useful first step in classifying software is to distinguish between *amateur* and *professional*.

The word 'amateur' is not used here in any derogatory sense. On the contrary, as I will show shortly, amateur software is a staple part of workaday computer use in education—or should be.

Essentially, amateur software is written by the user, who is sometimes the teacher, sometimes the student. There is no need for it to attain sparkling technical standards. It is enough that it should have a reasonable capacity to serve its immediate objectives in its own home surroundings. Indeed, so far from having need of high refinement and the smarter underlying qualities that are proper to professional software, such attributes can actually be undesirable where the time an amateur can spend in trying to provide them might more usefully be devoted to other things.

Amateur software does not ravel, and it should not try. In doing its work it should stay close to its origins and should be owner driven. In fact, as a rule, it is best it should be marker driven. If the owner is not the maker then at least they ought to be good friends.

Professional software is the kind you buy in the commercial market or obtain from some official source such as an education authority, a state-sponsored project or Santa Claus. The standards and qualities which respectable professional software should attain are altogether a different kettle of fish from those which apply to the amateur. Alas, there is currently (1985) still a large quantity of supposedly professional software on offer which falls lamentably short of these standards. What these standards are and why they are beyond the scope of most wise amateurs will be explained shortly.

Between amateur software and professional software there is an intermediate kind which we may term quasi-professional software. This is amateur software masquerading as professional. There is a lot of it about, and very dangerous it is too. It is not always easy to spot without going to the lengths of buying it and trying it. However, it is a safe bet that a program belongs to this category if it is ungainly, or difficult or frustrating to use, or if it has irritating quirks or is wanting in accompanying documentation, or it simply does not work properly. Apart from being intrinsically bad for the welfare of those who try to make serious use of such programs they can get computing itself a bad name, and may already have contributed to the expensive disappointments experienced by a number of publishers who have made attempts to sell 'educational software' in the consumer market. The technical and stylistic qualities are not the only aspects which matter. If the content is unsound, too, the potential for doing educational damage becomes stupendous.

The abundance of quasi-professional software is not surprising, for all you need is a home computer and you can set yourself up in business as a software house. It is one of the easiest ways of turning a hobby into a paying concern, or even possibly making a fortune, if some popular stories are to believed; so it is hardly a wonder that a couple of years ago droves of self-taught programmers were trying their luck in the marketplace like adventurers rushing to the Klondyke. Some of them are still at it.

This is not to say that all amateur-turned-professional software is shoddy. There are some examples which are truly excellent, but they are comparatively rare.

If you are a talented amateur, think twice before leaping into the role of professional. Your good ideas might well be made the basis of professional programs but it is probably better to get them professionalized by an organization suitably equipped to do so and which has a respectable track record than by changing your occupation.

AMATEUR SOFTWARE

Amateur CAL software can be divided into:

Plasticine programs
ULPs
Homebrew programs

Plasticine programs are programs which, like objects modelled from that substance in the infants' class, provide their educational effect in the making of them rather than in the final product. Once the job has been done so, too, has the educational part of it. For example, students who succeed in writing a program which will deliver the roots of a quadratic equation in response to input of the coefficients, distinguishing between the possible cases, are likely thereby to gain as great a mastery and as deep an insight into the topic as ever they would be the conventional route of working through exercises; moreover they will probably enjoy it a good deal more and will have experienced the added advantage of having to exercise self-reliance and self-criticism. Again, just about every program which children write in LOGO (real or imitation) is a plasticine program, while LOGO itself is a prime example of what may be said to be a plasticine language.

ULPs are 'useful little programs'. These are short programs, for the most part probably in BASIC, which can serve a valuable educational purpose, often out of all proportion to their diminutiveness. For example, a ten-line program which in converting a vulgar fraction to recurring decimal rattles off the complete digit cycle three times in full can be used as the source of a mathematics discovery project. Or a program which jumbles the letters of some chosen words, though trivial to code, can be put to happy effect in the junior classroom. Such programs can be written from scratch by teacher or by pupil, often with only a rudimentary knowledge of programming, or can be copied from books or magazines which increasingly contain listings of short programs of this kind.

Homebrew programs are more substantial programs which a teacher might develop over a number of evenings or, if very ambitious, might spend some weeks on, or even months (or get a bright pupil to do over the weekend!). The outcome of such labours could be something grandiose, like a simulation of the problems of organizing a bus service, or a puzzle sequence culminating in a derivation of the Lorentz transformation, or something merely for the moment such as a tailor-made program to help a particular child with some specific remedial problem. Before embarking on a big venture it is always wise to see first if a satisfactory professional version is obtainable.

Should teachers be able to program? Or, indeed, is it necessary for teachers to know anything about computers at all? It may now be taken as recognized that in any modern educational establishment there should be at least somebody who is competently informed about computing. The time has arrived when every teacher ought to be capable of making good use of a computer whenever it may be gainful to do so. There should be some understanding both of software and of hardware. What is needed is not demanding. Of the hardware, as with other items of educational technology, a teacher should know enough to be able to operate it, look after it and apply simple first aid, a competence similar to that which most people have concerning their care or tape recorder.

What programming language should teachers use? Apart from any other which they may have occasion to need, e.g. LOGO, the one essential language is BASIC. This has nothing to do with whether or not BASIC encourages slovenly habits, is wanting in structure or causes purists to tear their hair. BASIC is the *lingua franca* installed as standard in the actual machines which are available for use in schools and homes today. It is likely to remain so at least until the end of the decade. When at some future time the machines in common use have become multiprocessor, 32 bit, 80 MHz marvels and the popular on-board language is SPIZZO, then that will be the language which teachers will need to be acquainted with. For the present, whether we like it or not, the practical need is for a working knowledge of BASIC. The programming capabilities a teacher needs are comparable to those of being able to prepare a workcard—every teacher should be able to do that—rather than being able to write a textbook, which is accomplished by only a few. The necessary knowledge and skill can be taught from the beginning in a matter of three days. This should be a standard component of teacher training.

It is one thing to be able to program and another to be able to devise an algorithm. A teacher may conceive a program and know how to shape it and what the routines have to do, but might not so easily know how to make them do it. The devising of some routines calls for mathematics or technical intricacies which may reasonably be beyond the non-specialist programmer. The teacher needs a source of ready-made routines. Sometimes these can be lifted from other programs or copied from various publications. There is scope here for someone to prepare and issue a collection of routines for easy use by ULP and homebrew programmers in some suitable form. A further idea which might prove rewarding to explore is that of the kit of parts for a program, possibly provided on a disc which is furnished also with some utility programs for demonstrating or editing the material, which the user can put together according to personal style and taste. It would be something like making a cake with the aid of a packet of ready-prepared ingredients—cakemix software, you might call it.

PROFESSIONAL SOFTWARE

If you pay good money for a piece of software you should have a right to expect it to perform impeccably, to be polished in design, style and presentation, to be easy and pleasant to use, to be forgiving of a user's mistakes and frailties, to be exceedingly robust and totally reliable. To determine how near a sample of software might come to these requirements is not easy. Short of acquiring it and subjecting it to fierce and prolonged trialling you can never be sure unless it has already successfully passed such testing by others and gained a good reputation for itself. Be warned that one way you should never judge the true merits of a piece of software is from watching a demonstration of it. Deftly presented, a quite primitive program—or even a bogus one—can be made to appear superb. The maker of a homebrew program can often do the same, for he or she knows not only just what to type or press at any given moment but also what not to do. If something has to be input at some point the maker will know whether it has to be a number of a word and if, say, a number, he or she will know what sort of size it should be (these things do not always matter, but very, very often they do). But merely watching you will not discover (a) if you had not known what sort of size to enter how difficult it would be to find out or (b) if you had entered something unsuitable whether the program would 'go up in smoke'. If the program is homebrew and kept for use honestly as such—at its place of origin—these things do not matter because its creator, who has the advantage of possessing insight into the guts of the program, is on the spot to manage its running with, if needs be, a touch of manual override.

What does it take to produce a program of professional quality? Given the time and expertise needed to produce a simple homebrew load-and-go program (i.e. one which once loaded into the computer can be run without ever needing further access to discdrive, printer or any other peripheral) then the time and expertise needed to produce a professional version of the same thing would be at least an order of magnitude greater or, ignoring the question of how one might measure the expertise, ten times as much. Unlike the amateur version, the professional one must have professional design and style, must have watertight error handling, all inputs fully checked for validation, and most likely it should have inbuilt flexibility, i.e. a facility whereby the user can vary the mode or level at which the program will run. With homebrew this is unnecessary since the user can simply alter the program to suit the occasion. With a sophisticated piece of professional software this is out of the question. For one thing it might be written in a programming language the user does not know (e.g. it might be in machine code) and for another it might have critical attributes of length or memory usage which make tampering with it other than by an expert liable to destroy it. To provide an equivalent facility—enabling the user to modify the program's mode of action—in the professional version means providing a lot of extra code specially for the purpose. In addition, professional software should contain an ever-present means whereby the user can effortlessly summon guidance, clear and immediate, as to what choices are open to him or her and, having chosen, how to proceed. Thus the actual quantity of programming has to be considerably greater in the professional case. Moreover, the professional program has to run satisfactorily not just on the system on which it was implemented by the author but also on a majority of systems ostensibly of the same model. That may prove a headache, for it can happen that a program which runs happily on machines which came from the factory in August refuses to work properly on the batch of machines which it produced in July. This is because manufacturers often make little changes to their product as time goes by. A further cause for more work in producing a professional educational program is that it needs to be suitable for use not just with the particular pupils in the particular circumstances which the author was catering for but with a variety which may be spread over the country. Thus trialling the software becomes not just a matter of doing it on the spot in a single school or class but of checking that it is right for a far more extensive sample.

When it comes to a piece of software that has provision to access peripherals—e.g. loading or updating files on disc or outputting to a printer—the problem is immeasurably greater. The software has to cater for such possibilities as the wrong disc being in place, or being write-protected when the program tries to write to it, or the discdrive door being open—in fact to cater for sundry error conditions without spoiling anything for the user. The program must be able to recover from any such occurrence without turning a hair. If it has to be able to communicate with a printer the scale of the problem is greater still, for on the market there are several hundred different models of printer and most have little switches on them enabling them to be configured in a variety of different ways. This says nothing of the fact that there is also a variety of different possible interfaces and 'ports' or 'slots' through which the connections might be made. Needless to say, if the software is arranged to communicate satisfactorily with one printer setup there is no guarantee that it will do so with another. If it is homebrew software you have only to write it so that it does what you want with the particular equipment you are writing for. If it is professional, however, it will be expected to be able to cope with any printer it comes across. This is a tall order, but it can be done. It takes time, though, and more than ordinary know-how. Then, on top of the programming itself, the professional package needs much fuller documentation than any amateur software. In the case of ULPs or homebrew, if you need advice you can always ask for it direct from the 'horse's mouth'. On the other hand, if it has come to you from afar you may have to depend on the accompanying literature. It is fair to add, however, that software which is unusable without close study of hundreds of pages of user manual is an abomination and ought to be quite unnecessary if the software is made decently fit for human consumption in the first place. The printed guidance which enables you to run and operate the software should need to occupy little more than a single page. From there on the software itself should afford you all the necessary guidance and reminders as you go along. However, an (optional) introductory guided tour through the software is a desirable documentation feature, as is a technical reference section—something to refer to rather than to read solidly. In educational software the bulk of the documentation should be about the content of the program and perhaps about ideas on how to use it to felicitous educational effect.

Real professional software is exacting to produce and takes a great deal of designer power, programmer power, technical resources, time and money. Of these items the one currently in short supply is money. The total expenditure on software by UK schools in 1984 was around 2 million pounds, which barely covered costs of marketing, publication and distribution. Compare this with 80 million pounds spent in the same period on books (and this was considered lean fare). The cost of adequately responsible software development was able to be met by publishers only in rare instances. Some software development has been subsidized directly or indirectly by government: directly, for instance, by the now-expiring Microelectronics in Education Program; indirectly through the facilitation of software development by persons based and maintained within various educational establishments. The direct sponsoring of software is not always successful because an excessive proportion of the funding is all too often squandered on the creation of programs which properly belong to the amateur category.

There are many ways to go about creating professional CAL software packages but essentially they entail four main phases: origination of the idea, design, programming and trialling. There is usually a considerable degree of cross-feeding between these operations and the programming involved may be enormously more than ends up in the final product. The idea itself might be developed using primitive programs while the design stage may make use of programmed mockups to test the look and feel of proposed components of the package. The idea stage should be under the control of a teacher, or someone comparably competent, for it is here that the educational content and the pedagogy are first prescribed. However, it should not be just any teacher. It must be a teacher who not only has a flair for invention but whose work and outlook accord with the principle that learning should be enjoyable. It is strange that this needs to be said but it is a gruesome fact that there are many teachers loose even today who believe that to berate, punish and humiliate are necessary ingredients of educational practice. No hint of that kind of attitude should be allowed within a hundred miles of the birth of any educational software. Probably the most crucial part of the software development is the design stage. If this is done correctly the trialling will go through easily without early work needing to be unpicked to accommodate radical modifications. The ideal designer would be someone who all at once is a graphics artist, an uncommonly good teacher, a typographer, a photographer, a movie maker, a musician, a psychologist, a showman, a poet, a conjuror; he or she should love and understand computers and their funny little ways, yet should have empathy with those who fear or hate computers; and he or she should be able to spell. The main programming should be done by *bona fide* experts. Bugs in the end result should be virtually impossible. Trialling should take account of all the possible circumstances, locations, purposes, preferences, abilities and needs for which the package is intended to cater.

A good quality professional CAL software package can take three or four years to produce. Even the smallest may take nine months or more. A program which is a variation on an already established theme can usually be made more quickly than the original, and the use of software engineering and other techniques may speed things up in the future; but, for the moment, the process has generally to be reckoned a long and costly haul.

Artificial Intelligence for Society
Edited by K. S. Gill
© 1986 John Wiley & Sons Ltd

21. WHAT WILL AN INTERACTIVE VIDEO LOOK LIKE?

ANGUS DOULTON National Interactive Video Centre, London

Someone recently said to me that only the British would do something as blindingly just, fair and totally even-handed as setting up a centre to develop modern technology and then appointing someone to run it whose degree is in English language and literature.

Some things can be said straight away. Interactive video is going to happen this time. As it happens and as our work develops at the NIVC (National Interactive Video Centre), we are going to prove the immense power of visual images—and particularly of moving images—in the learning process and at all levels of learning. To me this is a truism that has largely been taken for granted but actually often ignored in our generation. It was known in my view to the artisans who decorated the roofs of Italian churches and in our own or almost our own time, it was more than well known to Einstein and a few others who simply did not have the technology to implement what they suspected. As people start to create programmes that consist of very short clips of moving video, controlled and directed by powerful computers, we are going to need to start reexploring the grammar of visual imagery. We are going to find that the computer is not the main agent but simply a partner in a situation in which the visual image is quite as important as the computer's undoubted power.

The next definite thing is that interactive video is going to be important because it will get closer than any other part of modern technology to ordinary people. It is already doing this through point-of-sale, public information (possibly entertainment) and certainly training. Interactive video is going to happen because there are strong commercial reasons for it to do so. The question for anyone involved in academic organizations, or even in quasi-academic quangos such as the NIVC, is: shall we let it happen or shall we try to pick it up and use it and develop it into an immensely powerful tool it could be?

The NIVC has been running since October 1984 and in that time we have seen something like 700 organizations involved in education and training. We have also seen the beginnings of IV production in the United Kingdom. As we have moved from meeting a simple demand of demonstrating interactive video towards suggesting that we should try to do something about the state of the art, we have formed three and a half different user groups. The first of these groups is the BIVA, British Interactive Video Association, which is an association of commercial production companies; the second is a group of educationalists who are concerned with developing IV in schools; and the third is the UK IV User Group, which is comprised of senior managers from industry, commerce and government who wish to work together. The half is a subgroup of UKIVUG, which comprises ten organizations which have joined together to commission development activity.

To me, these groups have one common objective, namely developing methods of creating and applying interactive programmes quickly, cheaply (within reason) and well. It therefore seems to me that they need to solve two problems: how interactive can we make it; how intelligent should we make it?

The point about interactivity seems to be that it is offering a chance to get closer and closer to the way most people think and, I suspect, learn. As it does so it is going to start challenging some of the rigid dogma of machine intelligence. It is not a bad vehicle for introducing humour. As applications develop, I believe that it may quite quickly get away from the rather corny heavy-handed humour being applied at present towards much more subtle applications of wit.

It is very definitely a good vehicle for presenting choices to people and I believe that it is the first technology that is well able to show the consequences of wrong choices—and there is a whole can of worms involved in persuading teachers and trainers that showing someone doing something wrong might actually be a good teaching practice.

Involved in making programmes interactive are some quite big questions about analysing how people learn and then applying the findings to producing subtle, flexible and quick programmes. As these develop we are bound to lead towards the next question: how intelligent can we make all this? This seems to me to be where you ought to be trying to link IV with expert systems, with IKBS (Intelligent Knowledge Based Systems) and with other forms of artificial intelligence.

To me, interactive video will not be intelligent in any real sense until it can start responding to individuals, can start not only to pick up the response to the stimulus but begin to guide the choices, can hold visual databases that can be accessed by open-ended shells and can help to cause profitable interactions between learners, tutors and knowledge bases.

You will notice that I have not yet spoken about videodiscs. It must be easy to conceive of everything I have discussed happening on a far larger scale. Very large visual databanks (hopefully carefully constructed), held centrally, accessed by multiple users down cables or satellites seem a distinct probability. And they will be very powerful. However, they will also be relatively useless if we have not done the work that is necessary now to find out far more about how people learn with this kind of technology, to develop properly responsive programming languages and to work out what 'interactive' really might mean.

From what I have been saying there are one or two things I would like to pick up. First, 'computer literacy' seems a very limited concept. The only form of computer literacy required is knowing that there are some tools to use. Second, I wonder how useful the concept of an intelligent tutor is.

I would like to end by saying that it seems to me that the most intelligent thing we can do is to choose to use these machines for proper purposes. The great danger in AI and IV developments is that we try to make the machines do things that they can do only with difficulty or not at all, and, in particular, that we allow the immense data storage capabilities of these machines to blind us to the fact that data storage by itself is not enough. Therefore the most important social job we can do now is to identity proper uses of technology and of people and to find and develop public applications.

Artificial Intelligence for Society
Edited by K. S. Gill
© 1986 John Wiley & Sons Ltd

22. INTERACTIVE VIDEO IN EDUCATION AND TRAINING

RAVI SINGH Black Rod Interactive Services, London

INTRODUCTION TO INTERACTIVE VIDEO

Both video and computers have a relatively long and chequered history in education and training. Although many good examples of educational video have been produced, their effectiveness is limited by the fact that they are linear and passive. That is, they do not actively involve the trainee in the learning process. Although with a video, a learner can review any particular sequence repeatedly, it is still a passive medium and cannot respond to an individual's learning requirements.

That computer-based training (CBT) has had slow acceptance can be partly attributed to the poor quality of some of the early courseware. In the absence of good and easily produced high resolution graphics and animation, such courseware relied excessively on text and lacked realism. A redeeming feature of CBT courseware is the incorporation of interactivity. A trainee is forced to interact with a lesson and proceeds through a network of knowledge, branching at each point of interaction and, therefore, receiving individualized instruction to a degree determined by the author of the courseware.

The convergence of computer and video technology to produce interactive video systems has meant that CBT is no longer restricted to text, and video need not be linear and passive. This integration of two quite diverse technologies has produced interactive video—a completely new training resource.

Interactive video is considerably different from both its constituent parts, in the manner in which the image on the screen is controlled and the flexibility it brings to the designer of training materials. A lesson based on interactive video may consist of moving pictures (e.g. archived documentary, drama material), still pictures (e.g. slides, photographs, paintings, maps), computer graphics and text. This mixture of media motivates the learner and contributes towards the effectiveness of the training resource.

Interactive video can be achieved by interfacing videotape or videodisc players to computers. The basic system is likely to consist of a microcomputer, a television or monitor, a videodisc/tape player and an interface which integrates the functioning of a computer, video player and authoring system (Fig. 1). Alternatively, a genlock system can be introduced which would provide the extra facility of text overlay (Fig. 2). A genlock is a device which will synchronize text/graphics output of a computer to a standard interlaced 625 line composite video signal from a videodisc/tape player. Such a system is capable of providing interactive branching because of rapid access and electronic addressing to video sequences. The use of such a system makes it possible to deliver individualized interactive audio-visual instruction.

THE IMPORTANCE OF INTERACTIVITY

We will consider the role of interaction in the learning process. Although educational psychologists have propounded several theories to explain the underlying psychological phenomenon necessary for learning to take place, few would dispute that learning is facilitated by the student actively participating in the learning process.

In passive learning situations, whether they are based on linear video programs, distance learning broadcast television or paper-based learning resources, the learner is not generally required to actively respond. The lessons' structure, pace, sequence and selection of material always remains the same. The decisions concerning the above are determined by the author at the time of the production of the training material and rarely are changes made thereafter. Such materials cannot take account of the differences in responses that individual users will make.

On the other hand, the sequence and selection of material in any interactive training program is determined by the trainee's response. In other words, an interactive system, by questioning the student, is able to both provide and receive feedback. That is, a dialogue is set up between learners and the delivery system which enables active participation by learners in their own individual learning experiences. The use of inter-

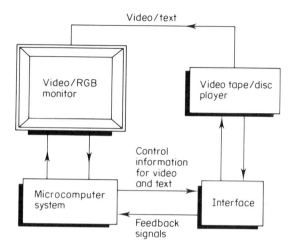

Fig. 1   A basic interactive video system

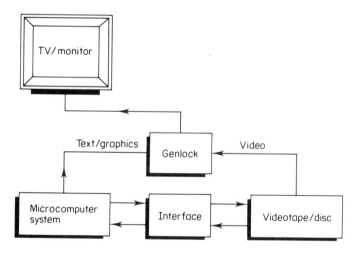

Fig. 2   An alternative basic interactive video system

active video has made it possible to further enhance such individualized and active learning experiences.

It is important to realize that interactive video is not the only learning medium which actively involves the student. One can successfully argue that some of the better designed textbooks and paper-based self-instructional material have been made more active by including in them activities such as problems to solve and workbooks to use. The degree of activity in these traditional learning resources can vary from slight to extensive. Indeed, interactivity is a design feature which can, with a varying degree of effectiveness and ease, be incorporated in most learning media. Such multimedia learning materials can be used by the trainer to provide effective and efficient training. In addition, a computer is included for the purpose of controlling the flow of information and facilitating a dialogue with the trainee. The resultant system has the potential for very high interactivity. The realization of this potential is dependent principally on two factors: the design of the courseware and the hardware employed to implement the design. Different combinations of these two factors can result in very different levels of interactivity.

TAPE VERSUS DISC

As stated earlier, an important way in which interactive video differs from linear video is in the manner in which the image on the screen is controlled. In IV the image which appears on the screen can originate from a video source, a computer or a combination of the two sources. The various hardware combinations available to implement interactive video differ from each other in the way that the two sources of image are used.

It is useful to explore the properties of the images that the two sources are capable of producing. Any image can be seen as consisting principally of three components (i.e. text, sound and pictures). The computer and the video source are both capable of storing all three components of an image, though the quality with which they are reproduced varies a great deal. Obviously, video is more suited to the storage and reproduction of pictorial and sonic images, and the computer is better for the storage of frames of textual material. Disc is capable of better video and sound reproduction when compared with some tape formats.

The adaptability of tape/disc-based video systems for use within interactive training systems is worth exploring. It is often mistakenly assumed that really effective interactive video can only be achieved by connecting a videodisc player to a computer. On the contrary, many of the features commonly associated with videodisc can be achieved using carefully designed interactive videotape systems. It can be argued that in many applications interactive videotape is actually a more desirable medium. This may be for economic reasons or for reasons which are associated with the volatility of the subject matter. Where the training material changes fairly rapidly, the time scales involved in the videodisc projects are such that its use may be prohibitive.

Videotape and videodisc differ from each other in a variety of ways. These differences have an important bearing on the way that the media can be incorporated within interactive systems. A summary of the major differences between the two is presented below.

Videodisc is faster?

Videodisc is a direct access medium and, as such, video segments can rapidly be displayed. Videotape, on the other hand, is a serial medium and therefore somewhat slower in accessing the desired sequences.

The access time has an important bearing on certain training programs, e.g. those dealing with motor skills. For many other applications short access times are not crucial. Furthermore, well-designed tape programs can minimize search times so that they do not interfere with the effectiveness of the program.

Videodisc is more flexible?

The videodisc provides two audio channels which can be used independently of each other or not at all. To the courseware designer it provides greater flexibility in the use of audio. The additional audio channel can be used to provide alternative commentary or perhaps just music and effects. Since the better designed IV programs are likely to include scenes lasting no more than a few seconds, the role of audio commentary is somewhat reduced. In fact, for many applications, audio commentary may be quite unnecessary.

Videodisc is able to mix more easily?

An important feature of videodisc-based IV is the facility it provides to mix video with text. This is by far the most important benefit of using videodisc rather than videotape. The use of this facility makes it possible to add meaning to still or freeze frames by superimposing text on them.

Complex pictures can be labelled and the students' attention drawn to particular parts of the picture.

Videodisc is more accurate?

It is often suggested that the accessing of particular video sequences on tape is not accurate and that videodisc is frame accurate. There are, however, tape-based systems which are frame accurate and can precisely locate segments within a few frames. The accuracy of tape-based systems can be measured in terms of playtime, and interfaces are available on the market which can access segments of video within half a second of playtime. This is in comparison with the frame accuracy of the laser vision videodisc system, which plays at 25 frames per second.

A tape specially designed for interactive use should take into account the capabilities of the hardware. Short scenes can be interposed with fade-ins and fade-outs, thus obviating the need for frame accuracy.

Videodisc is more expensive?

The present state of the technology is such that videodisc is available only as a replay medium. The original material is initially recorded on a tape before it is transferred to disc. This extra stage between recording and replaying is not only time-consuming but also adds considerably to the project budget. The cost gap between the disc and tape replay systems is, however, closing rapidly, and is no longer such a crucial factor in the choice between disc or tape.

Another important point to consider when evaluating the economics of videodisc- and videotape-based training packages is that in order to ensure good quality replay on the videodisc the original material must be recorded on high-band videotape. This fact often precludes the use of existing in-house video facilities which are often equipped with low-band facilities. The additional costs of hiring high-band facilities can run into several thousands of pounds.

Summary

Table 1 summarizes the main features of tape and disc. As can be seen, although the costs are higher, videodisc does provide considerable advantages over tape. It has better picture quality, freeze/still frame capability and is a direct access medium.

It is important to realize that it is the design of the instructional program, rather than any particular technology, which results in effective and engaging training materials.

Table 1   A comparison of tape and disc

	Access	Speed	Frame location	Mix video and text	Cost	On-site preparation
Tape	Serial	Reasonable	Reasonable	Yes	Cheaper	Yes
Disc	Direct	Faster	Accurate	Yes	More expensive	No

CONCLUSION

Interactive video is a flexible and useful tool for the trainer. Our experience has led us to conclude, however, that while the cost of videodisc remains at its present level, it will always be beyond the reach of some organizations.

A low-cost alternative is to adapt from the extensive range of video-based training materials already available. This brings its own design problems, but can result in viable individualized programs.

PART 6
AI, IT—Applications

Artificial Intelligence for Society
Edited by K. S. Gill
© 1986 John Wiley & Sons Ltd

23. AI AND THEORIES OF SOCIAL SITUATIONS

SHARON WOOD Cognitive Studies Programme, University of Sussex, Brighton

ABSTRACT

Very little is known about the underlying causality of social situations; our theories generally lack predictive power regarding situations that arise as a result of our action or inaction in social settings. Through our work on the development of an interactive ('expert') system to advise trainee teachers on how to deal with different types of lesson situations, we have set about formalizing existing knowledge of the processes operating in a very specific social setting: the classroom.

Through formal computational representation, we are able to gain a much greater understanding of the processes taking place and, in so doing, make accessible knowledge which is otherwise only to be found at great cost, in this instance, in the heads of experienced teachers. We also find that interaction between system and user for social domains is necessarily less prescriptive and thus less detrimental to individual autonomy in system use than is traditionally found.

INTRODUCTION

The Cognitive Studies Programme and Education Area at Sussex University are collaborating in the development of a Trainee Teacher Support

System (TTSS) for advising trainee teachers on problems they experience when taking lessons. We have adopted an expert Systems approach in representing the body of expert knowledge required in undertaking this task: that of the experienced teacher who is school-based tutor to these trainees. This knowledge is domain specific to the classroom and is passed on to inexperienced trainee teachers in consultative sessions away from the classroom.

A typical example resulted from a trainee's difficulties with teaching a humanities lesson when first beginning their teaching practice:

Trainee: I had problems starting the lesson . . . they want to know why they should do it.

Tutor: It's a good idea to start by recapping on previous lessons. Outline the course and tell them the aims of the course. Also, breaking down the lesson gives you confidence; knowing why you're doing what you're doing helps if you're worried about getting started. Relating aims and objectives to the lesson makes them concrete as well as abstract and helps you to stand up in front of the class. It can help in starting lessons if you can give them a reason why you're doing it; they may rather be doing something else, so you must explain aims and where the course is going. Pupils have high expectations in humanities for the lesson to be interesting. . . .

The work of the Expert Systems in Teacher Education (ESTE) Project is characteristically different, however, from that traditionally associated with expert systems. First, the domain of classroom teaching is essentially social and differs, therefore, from the typical areas of application for expert systems, such as medicine or 'fixing your automobile engine'. Our knowledge of what and why things happen in social situations is far less precise than it is for these essentially scientific areas and our theories about social situations tend to lack the predictive power usually associated with scientific theories. Consequently, what knowledge we have is difficult to formalize as rules of action or consequence in the format required for an expert system's knowledge base. It is not clear, therefore, to what extent it is possible to succeed in developing a system of this kind, although we might anticipate an increased understanding of the processes at work which underly the events we observe and experience in social situations (and of which we surely take account in interacting socially) through our attempts to formalize them within the system's knowledge base.

Second, we aim to be non-prescriptive in advising the system user; there are many ways to teach a lesson successfully, rather than a fixed number of solutions (although this is a contentious issue and there are those to be found who would strongly disagree with this stance[1]. Also, the Sussex PGCE course encourages the development of individual style in its trainee teachers, especially where this capitalizes on the strengths of the individual concerned and avoids their weaknesses. We prefer to view the user as an intelligent collaborator in identifying appropriate solutions to problems. The expert system, therefore, would not be the manufacturer of solutions to problems but the interpreter of problems in a non-superficial way, in order to provide information or explanation of a kind that will assist and guide the user in making decisions. It could therefore be called a 'decision support system'[2]).

It is probable that the lack of predictive power inherent in existing theories about social situations effectively precludes making advice prescriptive; nonetheless, it becomes clearer as time goes on that the position we have chosen to adopt on this may be the only acceptable one to users of expert systems (e.g. see Miller[3]).

PRELIMINARY WORK

The dialogue taking place between a trainee and the TTSS would be modelled on that which typically takes place between trainees and their tutor at the school where they receive their teaching practice. With this aim in mind, we conducted a study at a secondary school where five PGCE students were based, and to whom four tutors had been allocated, corresponding to the four subject areas of mathematics, biology, English and the humanities. We especially wanted to discover the range of topics on which trainees wanted advice, so that the system would ultimately be something they would find useful, although this study was confined to the initial phase of their training.

We found that trainees had a wide range of concerns, the greatest being the control of classes and how they would present themselves; whether the pupils would know they were 'only students', which might then again contribute to their problems of control. In response, tutors tended to stress the building up of a relationship with the pupils, initially at the cost of learning, with additional advice on avoiding confrontation, etc. For example:

Trainee: The pupils were very disruptive, so I told them they couldn't go until they had done the work.

Tutor: This is good, much better than saying they couldn't go until they were quiet. It is important to establish a relationship with the class at the cost of work, to begin with. Be consistent in your approach: rules, expectations, what they can get away with, and they will start relating to that.

Trainees also experienced problems in coping with classes of different ability, especially mixed ability classes, and similarly in getting their relationships right with different age levels. Advice from tutors ranged from such know-how as selecting 'markers' within the class by whom to judge the pace of work to general management advice. For example:

Trainee: I had problems keeping them awake; they were totally bored with it: some of them have computers at home and knew it all and some of them hadn't a clue.

Tutor: This is the classic problem of mixed ability teaching.

Trainee: This will cause problems with the project I was going to set; I will have to scrap it.

Tutor: How about splitting up the group and giving those at different levels different projects?

Trainees experienced difficulties in their general methods of communication. For instance, a tutor might draw attention to the aims of an activity such as questioning: if it was to discover the general range of knowledge a class already has on a topic, then one might not get a very accurate point of view by getting answers from just a few bright pupils at the front of the class who always have their hands up. They were encouraged to think in terms of the aims and objectives they had in mind when undertaking an activity with a class. Similarly, the need to explain things adequately was emphasized. For example:

Trainee: The kids grasped the idea quickly and got very excited but didn't seem to get very far with the written exercises in their textbooks.

Tutor: The sudden switch to the formality of laying out answers and using the terminology in the textbook can throw pupils; relate the notation to what they've done orally.

Trainees often experienced problems with simply organizing large groups of individuals and managing them effectively, so would seek and get advice on general management problems as well. For example:

Trainee: During the practical the pupils were shouting for me to come and see their slides. I was all over the place; in the end I said 'You'll just have to wait!'.

Tutor: Yes, just explain you're not superwoman. Explain you're going to start over here and work in this direction, or say 'I'll do you, you and then you'.

The interaction underlying the seeking and giving of advice which took place during these sessions appeared to be aimed at diagnosing specific difficulties raised by the trainees themselves. The tutor would follow this up with questions about the context surrounding the situation the trainee described: previous activities, previous classwork, aims of the trainee for the lesson, etc. Using their experience the tutors were then able to reconstruct the scenario the student described, making an accurate interpretation of the situation and responding accordingly with an appropriate explanation and advice.

In performing the task of tutoring, it is clear that tutors are bringing into play a wide range of knowledge, for instance, about the sort of things that can go wrong in lessons: not explaining things adequately; not questioning effectively; not allowing enough time, etc. They can distinguish between different classes of error: the error of having unrealistic expectations for a class or planned activity; or of not ensuring the class is in a mood appropriate to a planned activity—too sleepy last thing on a Friday afternoon to effectively work quietly through an exercise book, say, or too boisterous for a controlled class discussion without some calming activity such as reading quietly to themselves for five or ten minutes beforehand.

Tutors also need to be able to distinguish between good problem indicators and red herrings: knowing those aspects of an event which provide useful information about the nature of a problem, rather than spurious events. For instance, lessons can suddenly go wrong when going from one activity to another. Discovering a problem has occurred at such a transition point in a lesson provides useful information about the nature of that problem; it is unlikely that the class has *suddenly* become awkward and uncooperative, and perhaps points to the fact that the task has become too difficult for the class or that they are not clear about what they should be doing at this point in the lesson.

Tutors also require an understanding of the processes at work within the classroom which underlie manifest events. By recourse to their appreciation of these processes, tutor's are able to infer the probable antecedents to the events trainees describe. Without this their sole response would be to the superficial characteristics of a situation as it presents itself, without being able to direct comment towards those factors which contributed to the problem arising in the first place. For example:

Trainee: The first lesson went very well, but I had a struggle to keep them quiet and teach in the second lesson.

Tutor: When kids start to play up in lessons when previously they've been good, this can be because the work starts to get a bit beyond them and their whole pattern of behaviour can change.

Last, but not least, is the knowledge tutors possess specific to their task of tutoring: knowing how to best present advice to a student, how to set about obtaining information about their aims and objectives, and how to relate advice to the insights this gives the tutor about the knowledge possessed by the trainee and their psychology, so that the advice will be constructive and not destroy the trainee's confidence in being able to deal with the situation.

The tutor's knowledge is implicit, however, in the responses made to trainees, and requires in-depth investigation involving follow-up interviews, etc., to arrive at an understanding about what led a tutor to ask particular questions and to give specific advice. This is a difficult task: techniques available are generally problematic[4] and not helped by the tendency when reflecting upon events to rationalize decisions which were made.[5]

We see hints of processes underlying the transformation of one classroom state into another in some of the responses given by tutors to various incidents. For example:

Trainee: As I was talking, the group started to split off into small groups which were talking amongst themselves.

Tutor: The attention span of the fourth year CSE group is not long enough for a whole lesson based around a talk; if the standard is low, you must go quicker to keep attention and relate the content to the pupils more.

Following up tutoring sessions with interviews should also provide further insights into processes operating here, although a similar process to the one which would seem to underly the above example has been described in the literature: a study conducted by Kounin[6] identified a process of 'satiation'. He measured this by observing levels of 'work involvement' and 'freedom from deviancy' in class pupil's behaviour: the more satiated the class became the less they demonstrated evidence of work involvement and freedom from deviancy. Successful teachers stemmed this process by the interjection of challenging remarks like: 'you'll really need your thinking caps on for this one', or statements about the enjoyability of the task like: 'I know you're going to like doing this', and their own enthusiasm for the activity. It appeared that these motivating statements reduced the rate of satiation. It is possible that 'satiation' and the 'motivation' derived from the teacher's remarks are functionally interrelated, so that motivation alters the nature of the satiation process itself.

To enable the system to apply the tutor's knowledge to a problem we must formalize the descriptions of the incidents trainees describe. We do this by defining a vocabulary of terminology that corresponds to the way in which teacher-tutors conceptualize these situations. We have consulted various sources to obtain an appropriate vocabulary: the literature on classroom and teacher training research (e.g. Wragg[7]); the observation study we undertook; and additional studies with experienced teachers seconded to Sussex courses. These teachers were shown videos of trainee teachers taking lessons, which we hoped would stimulate them to think explicitly about those aspects of the classroom and lesson which they felt trainees in their experience were failing to take account of. The descriptors which experienced teachers use are good indicators of the way in which they conceptualize situations, and how they are applied gives useful insights into how they distinguish between characteristically different situations which to the untrained observer might appear very similar.

For instance, we might consider a class which was essentially 'out of control' (not doing as the teacher would wish) to be also uncooperative. However, in our observation study the following situation arose. A class had been split up into groups to discuss their arguments for why they should be allowed to do what they wanted with a piece of land. Each group represented a different interest group: business men who wanted to build a supermarket; a community group that wanted to build a children's playground; etc. The groups would then take turns to present their arguments to the rest of the class. The lesson broke down, however, when it came to the presentations because the rest of the class would not stop talking. The class had become 'uncontrolled' (one descriptor). Yet they could not be described as 'uncooperative' (another descriptor): they were, in fact, enjoying the lesson and enthusiastically taking part. The tutor perceived the problem as one in which the class were unaware of the change in what was expected of them, through lack of cues following the transition from one phase of the lesson to the other.

We found it necessary therefore to define the meaning of a descriptor (word in our vocabulary) about the class' behaviour, partly in terms of their beliefs, in order to capture these distinctions which the tutor clearly perceived. One might, for example, describe a class as 'participating' (third descriptor) by describing the task appropriateness of their behaviour according to what they believe the current task to be, whereas an uncooperative class would be one whose behaviour was *not* appropriate to what they believed the task to be. However, under either set of circumstances, the class is *uncontrolled* if their behaviour is not task appropriate to what the *teacher* believes to be the task. The discriminatory power underlying the terms teachers use in perceiving and conceptualizing classroom situations is the basis for the tutor making the appropriate response to the trainee in a consultation.

SUMMARY

There is a body of domain-specific expertise belonging to teacher-tutors in undertaking their tutoring task which may be captured in an expert systems representation. This consists of: factual knowledge about the things that can go wrong in lessons and different classes of problems; strategic knowledge about what information is useful in diagnosing problems and evaluating the significance of what the trainee reports; deep knowledge about the processes which underlie the transition from one classroom state to another; and meta-knowledge or control knowledge about managing consultation sessions with trainees and making the best job of teacher-tutoring.

Our observation study revealed the nature of the dialogue which takes place between trainee teachers and their teacher-tutors to be essentially interpretative and explanatory, dealing with specific problems. Further studies involving interviews with tutors will help to reveal the knowledge implicit in their responses to trainees' problems. This will also enable continued formalization of the descriptions of situations as conceptualized by experienced teachers in terms appropriate to their underlying model of the classroom. As this work progresses, we will also be undertaking the implementation of a prototype system which will incorporate the knowledge we have obtained to date and inform us on our approach to representing classroom situations.

This work presents an opportunity to formalize what teachers know and to extend it, and in so doing gain a greater understanding of the processes we are investigating. Theoretical growth increases predictive power, informing our action—whether at the level of policy, strategy or tactics, in this case in the classroom—and affording us greater control over intervention in these areas, when seeking to advise.

Those who benefit gain not only from a theoretical insight and its consequence but also, eventually, we hope, the greater accessibility to information which at present, being in the heads of teachers, is a costly and consequently restricted resource. In making available this resource, we value an approach which does not reduce the autonomy of those seeking advice by excluding them from the decision-making process.

REFERENCES

1. Lacey, C. (1977). *The Socialisation of Teachers*, Methuen, London.
2. Freyenfeld, W. A. (1984). *Decision Support Systems*, NCC Publications, Oxford.
3. Miller, P. L. (1984). *A Critiquing Approach to Expert Computer Advice: ATTENDING*, Pitman Publishing Ltd, London.
4. Gammack, J. G., and Young, R. M. (1984). 'Psychological techniques for eliciting expert knowledge'. In *Research and Development in Expert Systems* (Ed. M. Bramer), Cambridge University Press, Cambridge.
5. Calderhead, J. (1984). *Teachers' Classroom Decision Making*, Holt, Rhinehart and Winston, London, p. 4.
6. Kounin, J. S. (1970). *Discipline and Group Management in the Classroom*. Holt, Rinehart and Winston, London.
7. Wragg, E. C. (1974). *Classroom Teaching Skills: the Research Findings of the Teacher Education Project*. Nicols, New York.
8. Eggleston, J. (1985). 'Subject centred and school based teacher training in the postgraduate certificate of education'. In *Rethinking Teacher Education*, (Eds D. Hopkins and K. Reid), Croom Helm, Beckenham, Kent.

Artificial Intelligence for Society
Edited by K. S. Gill
© 1986 John Wiley & Sons Ltd

24.

LANGUAGE TEACHING, COMMUNICATION AND IKBS

STEPHEN GWYN JONES SEAKE Centre, Brighton Polytechnic

ABSTRACT

There is a considerable shortage of computer software designed to teach English to people who are learning it as a foreign language. There is an even greater paucity of such material aimed specifically at adults.

It is suggested that effective computer-based material for teaching English to adult EFL students can only be produced by considering three factors:

1. The particular needs of adult students. Adult students prefer material which relates to the real world. They tend to dismiss games as being more suitable for children, and learn better when dealing with topics which have direct relevance to them.
2. The methods currently being used by language teachers. One of the more modern approaches to language teaching stresses a functional approach to language, which advocates teaching language as communication, rather than concentrating entirely on syntax and grammar.

3. The means by which computer-aided learning (CAL) techniques
   can best be used to meet the needs of both staff and students. The
   purpose of many drill-and-practice programs could be
   accomplished just as easily by the use of pen and paper. A logical
   way to use the computer to teach language would be to devise a
   system which could answer questions about a topic of interest to
   the student.

A system is described which works on a microcomputer and which
can give advice about diet and healthy eating habits in response to questions asked in a limited subset of natural language.

The programs which are described in this paper were written as part
of a two-year project conducted jointly by the SEAKE Centre at Brighton
Polytechnic, which carries out research into the learning difficulties of
children and adults, and the Friends Centre, which is an adult education
establishment in Brighton. The project was financed by the EEC Social
Fund, and its purpose was to discover ways in which computer-aided
learning techniques could best be used with adult students learning English
as a second language.

There is a considerable shortage of software designed for second
language teaching. One reason for this could be the difficulties which
teachers and programmers often experience in expressing their ideas to
each other. Higgins and Johns state that most programmers are unaware
of the changes that have taken place in the modern language classroom
in the last twenty years,[1] and it is true that most of the programs currently
available are of the drill-and-practice type, and concentrate heavily on
sentence structure at the expense of the actual communicative function.

A logical approach to designing software of this type would be to
consider three factors:

1. The needs of the students who would be using the system
2. The methods actually used by language teachers
3. The means by which the computer could be used to make a contribution
   to language teaching which it would be difficult to duplicate using
   another medium

Students from other cultures are generally more interested in teaching
material with which they can identify. Ideally this material should not be
culture specific, but should be relevant to as broad a cross-section of the
class as possible. They also want material that treats them as adults, and
for this reason many students are interested in using the computer for
purposes analogous to those for which it is used outside the classroom,
that is to say, for manipulating data and knowledge.

The subject of teaching methods is obviously a separate topic in its
own right and really requires a teacher to speak about it authoritatively,
but it is possible to say that one of the strongest influences on modern

language teaching is that of the communicative school, which sees language learning as the acquisition of the ability to communicate, rather than of the mastery of particular grammatical structures. Williams lists three characteristics of this approach:

> The major characteristics of Communicative Language Teaching appear to be three. They may be related to the areas of syllabus design, methodology and materials. At the level of syllabus design, the dominant feature is relevance to the learner's needs. At the level of methodology, the concern is with meaningful communication; at the level of material it is authenticity.[2]

This approach places emphasis on the function of language rather than its structure. Littlewood points out that from a structural point of view, a sentence like 'Why don't you close the door?' is unambiguously a question, but from a functional point of view it could be a complaint or an order, depending on the context.[3]

As regards the ways in which the computer can best be employed in language teaching, many language teachers object to drill-and-practice programs on the grounds that they could achieve the same results just as easily (and much more cheaply) by using a pen and paper. Given the criteria outlined above, one logical approach would be to write a program that could simulate a conversation with the student. The program that springs immediately to mind is ELIZA.[4] However, as anybody who has used programs of the ELIZA type knows, you only have to type in nonsense to make ELIZA talk nonsense. In order to write a program that would be able to have any kind of meaningful interaction with the student, it would be necessary to place restrictions on the area with which it deals. One way to do this would be to devise a system which has knowledge about a certain topic and to make it able to appear to hold a simple conversation on this topic. A program such as Carbonell's SCHOLAR[5] can do this, but it is a very large and complex program written for a mainframe computer. Moreover, SCHOLAR was designed to review the student's knowledge of the geography of South America, and this is not necessarily a topic that will appeal to the broadest possible cross-section of students. One of the main advantages to using a computer in this way is that if a topic of interest to the students is chosen, the interaction with the machine can be used as a basis for conversation in the class, which can help spoken language, as well as reading and writing skills. The use of a computer program as a catalyst for discussion in the classroom is documented by Nichol and Dean[6] in connection with the teaching of history. Kershaw suggests that the major application of computers in communicative language teaching is 'acting as a database to stimulate role play and discussion between learners, with or without a teacher present'.[7] This also takes some of the burden from the programmer, because it means that the program can be used as a basis for the teaching of aspects of language which the program itself could not handle.

The original topic chosen was that of diet, because most people enjoy eating and take some interest in what they eat.

Three programs were developed in the course of the project. The first of these was written in BASIC and dealt not with diet itself but with cooking and recipes. The reason for this was that cooking was considered to be a subject that was easier to define than diet. The system prompted the user with such questions as the name of the meal, number and type of ingredients, the country the meal came from, method of preparation, and so on. This information could be accessed by a very simple pattern-matching procedure so that a sentence like 'Tell me about chili con carne' or 'What meals use cheese?' would produce the appropriate response from the database. The main disadvantage of this program was that BASIC was not the best language to use. A considerable amount of work has to be done to approximate to the kind of facilities that are built into a declarative language like PROLOG, or even a relational database like DBASE II.

The second program developed was a small and tentative prototype expert system for diet which was constructed using the Microexpert shell. This contained about thirty rules of the form: IF X eats a lot of fatty food AND X doesn't get much exercise, THEN X should cut down on cholesterol (with a certainty factor of. . .).

There are two disadvantages to using an expert system shell of this type for language teaching:

1. The presentation format used by the shell is very rigid. It had been hoped to achieve a fairly relaxed, informal style of presentation, but there was a certain incongruity between the language of the shell and that of the rules, producing such anomalies as: 'The fact that YOU EAT A LOT OF STODGY CANTEEN FOOD will moderately strengthen the hypothesis that YOU NEED TO TRY TO EAT A MORE BALANCED DIET.'
2. Goal-driven systems tend to fire questions at the user in a way that is inappropriate for the purposes described previously. The user's only contribution is to enter either 'yes' or 'no', or a number between -5 and $+5$, to indicate a degree of certainty. For the purposes for which expert systems are normally used, there is no reason why this should be otherwise, but in the context of language teaching, many adult language students need to be encouraged to use their initiative in the classroom, as some students tend to take a rather passive approach to learning.

The most appropriate way to develop software of this type would probably be by using a programming environment such as APES rather than a shell, as this would give the programmer more control over the interaction.

The third program is an advice system written in MicroPROLOG for the Apple II, which can respond to a few simple questions about diet. These can be expressed in a fairly comfortable way in natural language, and the system can also volunteer information where this is appropriate.

The program can answer four types of question:

1. What nutrients does a particular item of food contain?
2. What is the best way of cooking a particular food?
3. What foods contain a particular vitamin or other type of nutrient?
4. What is the meaning of a particular word?

The system of pattern matching used breaks down the input sentence into phrases, and many different combinations of these can be used so that the same question can be expressed in many different ways.

Questions are evaluated by matching the object of the enquiry (e.g. tell me a good way to cook beef) to the generic food type to which it belongs, so that a question about the contents of soya beans will be interpreted as a query about the nutritional contents of pulses, and will elicit the response that, like most pulses, soya beans will probably contain protein, vitamins B and C, minerals and energy. It also happens that soya beans are exceptionally rich in both protein and iron, and since this fact is explicitly mentioned in the database, the system will also give a reply to this effect. *Ad hoc* information about pulses in general will also be mentioned: 'I know that if you mix pulses with cereals (like red beans and rice) this means that the protein will be absorbed better.'

If the system cannot immediately recognize the object of the query, but the question is phrased grammatically, a search is made for any name in the database which makes an approximate match with the name of the item of food mentioned; therefore if 'Tell me a good way to cook beaf' was typed in, the system would respond: 'Please excuse me while I look for the word "beaf" in my dictionary . . . (pause) When you said "beaf", were you thinking of beef?'

If a question is asked about a new item of food and the system can find no plausible match, the user is asked to indicate the generic food type to which the item belongs, and this enables the system to map it into its data structures. It can then answer any of the questions about that food item that would normally be within its capacity.

If the question is phrased ungrammatically, the system scans the input for keywords in an attempt to find out the intention of the query. If any of the keywords is recognized, the system displays examples of ways of phrasing that type of question to which it can respond.

Questions about cooking are answered first by finding the generic food type, as in the previous instance, and then by finding the ways in which that particular type of food is usually prepared. Thus poultry may be prepared by boiling, frying, grilling or roasting, but pulses are generally only prepared by boiling. However, it is a fact that if you boil any kind of food, some of the vitamins will be absorbed into the water, and this information will be given in connection with any food that is cooked by boiling, with all variables instantiated to the name of that particular type of food.

Questions about which foods contain a particular nutrient and about the meaning of words are answered by a straightforward lookup through the database.

Sometimes the explanation of words produces other words that can also be looked up. The question 'What are B vitamins?' produces the reply 'B vitamins are needed to protect against anaemia. Vegans may also need to take vitamin B tablets.' An immediate objection to this type of reply is that the explanation itself contains difficult words. This is intentional, as explanations are also available for such words as 'vegan' and 'anaemia'. In this way the student can be exposed to a wider variety of constructions than would be available from a straightforward explanation of one word.

The program has been piloted at the Friends Centre, an adult education centre in Brighton. Students have found it interesting to use and staff have agreed that it can be a useful teaching tool in the language classroom.

The scope of the program is fairly limited because it was written for the Apple II, but on a slightly more powerful machine it would be possible to extend the range of the parsing mechanism, so that incorrect grammar could be interpreted and errors could be diagnosed by the system, which could then provide appropriate feedback. This could be done by including parses of anticipated incorrect sentences as well as correct ones.

A further development might ideally be to design an authoring package, which would enable teachers to design their own programs around subjects of their own choice. Part of this could involve designing a simplified version of the type of grammar kit available in many mainframe implementations of PROLOG, whose use for language teaching has been documented by Kahn.[8]

Programs of this type, designed to deal with such topics as first aid, home safety, care of the elderly or living on a low income, would be valid teaching aids which could motivate students by encouraging them to gain knowledge useful to them on a personal level.

ACKNOWLEDGEMENTS

I am grateful to the following people for the encouragement and valuable criticism which they gave during the course of the research project described in this paper: Dr K. S. Gill of the SEAKE Centre, Brighton Polytechnic, Dr C. S. Mellish of the Cognitive Studies Programme, University of Sussex, Janet Price and John Traxler of the Friends Centre and Dr F. N. Teskey of the Department of Computing and Cybernetics, Brighton Polytechnic.

REFERENCES

1. Higgins, J., and Johns, T. (1984). *Computers in Language Learning*, Collins.
2. Williams, E. (1983). 'Communicative reading'. In *Perspectives in Communicative Language Teaching* (Eds K. Johnson and D. Porter), Academic Press.
3. Littlewood, W. (1981). *Communicative Language Teaching—An Introduction*, Cambridge University Press.
4. Weizenbaum, J. (1976). *Computer Power and Human Reason—From Judgement to Calculation*, Freeman.
5. Carbonnell, J. R. (1970). 'AI in CAI: an artificial intelligence approach to computer-aided instruction'. *IEEE Transactions on Man–Machine Systems*, No 4.
6. Nichol, J., and Dean, J. (1984). 'Computer-assisted learning in history'. In *New Horizons in Educational Computing* (Eds M. Yazdani and A. Narayanan), Ellis Horwood.
7. Kershaw, P. (1985). 'Mileage from the Micro—the computer in the modern language classroom'. *Modern Languages in Scotland*, January.
8. Kahn, K. M. (1983). 'A grammar kit in PROLOG'. Uppsala Programming Methodology Laboratory, Uppsala University, Sweden.

Artificial Intelligence for Society
Edited by K. S. Gill
© 1986 John Wiley & Sons Ltd

25. AN EXPERT SYSTEM ASSISTANT FOR HUMAN SERVICES PERSONNEL

M. J. WINFIELD, S. K. TOOLE, R. T. GRIFFIN and **P. M. DAVIES** City of Birmingham Polytechnic, Birmingham

INTRODUCTION

A number of expert systems have been developed over the last ten years, but they do not appear to have made much impact in the area of social work. In this paper we discuss the potential usefulness of such systems to social workers and some of the issues involved in designing an expert system for the domain of enuresis (bedwetting).

WHY DID WE CONSIDER AN EXPERT SYSTEM FOR USE BY SOCIAL WORKERS?

We believe that social workers, like many other professional workers, need as much assistance as possible if they are to make decisions which will be beneficial to all concerned—the client, the social worker or the social services department. This is in no way meant to decry the ability of social workers. We would argue that the vast majority of social workers are very professional and divisioned in the way they carry out their investigations. However, it is impossible for any one person to be capable of knowing all there is to know about the many different aspects of social work. Even a person considered to be an expert in one particular domain (area) of social work is still unlikely to know all there is to know about that domain. It is all too easy to miss (or forget to enquire about) an important piece of information, with potentially disastrous results.

It is our belief that the use of an expert system working in a particular domain could greatly assist a social worker to:

1. Ensure that all necessary aspects of the case have been considered.
2. Gain confidence in their approach to solving the case.
3. Seek advice on the course of action to take.
4. Seek advice when they are encountering difficulties.
5. Enable a social worker to safely train in an area with which they are unfamiliar.

You may well be saying: 'Yes, but could not a conventional piece of software have been used to assist the social worker?' The reply would have to be yes/no. Yes, if the problem domain is very well defined and there is a solution path which can always be followed. That is the problem that can be formulated algorithmically, and a known algorithm exists for the solution of such problems. However, in social work terms this is not normally the case. Social work, by its very nature, means that problems are not well defined and that they are not solved by simply slotting values into an algorithm. The solution path is generally unknown before working on the problem and consequently a solution involves the social worker applying a combination of good practice and heuristics (rules of thumb). We are therefore led to the conclusion that conventional computer systems are not really suitable for use in aiding social workers to solve a problem within a particular domain. They need more help than a conventional system can offer. Expert systems have a number of advantages which would appear to make them potentially very useful in the field of social work.

In particular, the following reasons lead us to consider expert systems as having a very important future:

1. No specific solution algorithm is supplied to an expert system. The system determines its own solution path by considering the facts currently known about a problem and then by drawing its own inferences.
2. Expert systems have an explanation facility which enables a user to query the advice suggested by the system. This enables the user to gain confidence in the result, although the final decision whether to accept or reject the advice is still left to the social worker.
3. The knowledge base can grow incrementally. This means that we can start off with a small system and allow it to grow without fear of having to rewrite it as the system grows.
4. Knowledge from several sources may be accumulated. The user poten-

tially has, therefore, a much wider access to knowledge than he/she possesses alone.

Selecting the problem Domain

The Department of Sociology and Applied Social Studies at Birmingham Polytechnic has for some time had an interest in the concept of expert systems and their potential use in social work. In particular, there is interest for their use in:

1. Giving advice in complex social work decision-making situations
2. Training social workers
3. Simulating the social work theoretical process in an attempt to aid understanding of the relevance to particular areas of practice

Why we Chose Enuresis as the Problem Domain

The problem domain to be selected had to be self-contained. That is, we needed a domain which would ensure that the system to be built would tackle a real-life problem. We did not want to fall into the trap of building a toy system which, whilst being easier to construct, would probably not show the true difficulty involved in building an expert system and would be open to severe criticism from professionals working out in the field who may argue that it was just a nice academic exercise. On the other hand, we did not want an area which would lead to a very large system being created, which would probably take several man-years to build. The result of our deliberations led us to choose enuresis, which on the surface may seem an unusual area. However, it was chosen for a number of reasons:

1. Enuresis is 'domain specific', i.e. it is in some ways a discrete area of intervention which suits the constraints of an expert system.
2. Enuresis is a common problem of which few social workers have an in-depth knowledge, yet it has a very high success rate for intervention. In addition, it is one of the most thoroughly researched areas of intervention with a large amount of developmental and empirical evidence at the outcome of specific methods of intervention.
3. One of the authors has an in-depth knowledge of this area and easy access to other 'experts' in this area.
4. The behavioural casework approach is particularly applicable to intervention in this area. It was assumed that using a systems framework and a behavioural approach would be most amenable to the process of 'knowledge engineering'.

The domain of enuresis therefore met the criteria of both the computer experts and the social worker to enable a prototype to be developed.

DESIGN OF A PROTOTYPE EXPERT SYSTEM

Knowledge engineering

It is recognized that the power of an expert system comes from its knowledge base, not its inferencing mechanism. This means that the process of collecting the domain-specific knowledge from the expert and the structure of the knowledge base (the knowledge engineering process) are critical to the production of a good expert system. Unfortunately, the knowledge engineering process was not as easy, nor as straightforward, as we had expected. The domain knowledge which the knowledge engineer is seeking to collect will include facts, beliefs and heuristics together with rules of inference which will guide the use of the knowledge. The need of heuristics is very important since the problems faced by an expert often do not have an easily formalized or algorithmic solution. The ideas which underline an approach to expert problem solving are shown in Table 1.

Table 1   Composition of knowledge

Knowledge = facts + beliefs + heuristics
Expert system solution = (knowledge + inference rules). (inference engine + explanation)

The knowledge engineer will normally collect the knowledge about a domain by interviewing one or preferably more domain experts. This was the process we used, initially relying upon the skill and experience of a single expert, but reinforcing his knowledge with the aid of papers in journals and books. The intention has always been to build the initial knowledge base by using a single expert and then to refine the knowledge base by consulting other experts.

We soon ran into difficulties, probably due to the fact that this was the first time any of the team members had either worked together as an interdisciplinary team or the first time many of them had experienced a knowledge engineering session. In addition to this the various members of the team had different agendas at the start of the project:

1. *Social work researcher*
   (a) To obtain more theoretical knowledge about expert systems
   (b) To test out the logical consistency of the behaviour casework method
   (c) To understand what the philosophy and psychology assumptions were of these so-called 'artificial intelligence' systems
   (d) To work as a stable team, e.g. computer researcher

2. *Computer researcher*
(a) To obtain 'pure knowledge' from the social work researcher in a form of rules
(b) To gain experience in this domain
(c) To work as an individual in a team of varying composition selecting according to task and training needs of the computer centre staff

Hindsight, providing as usual such clear wisdom, demonstrated that the first task should have been to build a team and set clear *objectives*. A variety of issues were raised which may be of value to others engaging in interdisciplinary work.

1. The computer expert must communicate exactly what they require. Examples are one of the best ways, particularly when the social work researcher may have predefined ideas about programming, etc., which must be relearnt with regard to expert systems.
2. It is necessary to clarify to all the team what is meant by a social work/ psychological theory. The confusing 'common sense' of the non-social workers caused much confusion.
3. The use of language must be extremely precise. The social work researcher used the same word for slightly different concepts and different words for the same thing.
4. The information must be passed and recorded in an agreed mutually understandable form.
5. Meetings must be controlled with clear aims and objectives set, and conclusions discussed before closing the meeting.
6. The exchange of literature on both the domain and particular aspects of the expert system followed by a 'seminar' is recommended.
7. Partial systems should be built early to test out the system and clarify team objectives.
8. Attention needs to be paid to group progress in addition to content, especially in interdisciplinary groups, as this seriously affects the group task performance.

However, with determination and persistence the team started to achieve results from the knowledge engineering sessions and approximately thirty rules were produced from five sessions.

In order to create an overall view of the knowledge, we found it useful, but certainly not essential, to represent the knowledge on paper in the form of a tree. This also enabled the domain expert to have a better understanding of the relationship between the items of knowledge supplied and also to understand the overall structure of the knowledge base. Obviously as the size of the knowledge base grows it becomes more difficult to structure the complete knowledge base in this form.

SOME CONSIDERATIONS FOR DESIGNING AN EXPERT SYSTEM

A prototype expert system has been developed using the microexpert systems shell. We have found that its design needs the following further considerations for its practical applicability.

Man–machine interaction

1. Use of certainty text (i.e. very likely) instead of certainty values (i.e. 3.5) is more appropriate.
2. Keep screen clear and concise with sufficient information to infer the choices the user has. Use plenty of clear screens and highlighting and even large text. Colour would also be useful but could not be used with the microexpert.
3. Explanations when asked for must be as detailed as possible. Levels of help are to be utilized for both levels of proficiency and different professional groups (social workers are not *au fait* with computers). Social work users could be working at a variety of levels of expertise.
4. System to give suggestions not order or commands, with good clear explanation facilities.
5. Output of messages—plenty of textual descriptions of suggestions.

Future considerations

1. No graphics facilities are available and it is felt that a future system will need computer-aided instruction with the aid of diagrams and pictures. This must be considered as relatively long-term research.
2. The system has no security mechanism. This is an essential item for any future system dealing with the personal details of clients.
3. A future system will need a mechanism for checking the consistency and integrity of the knowledge base, i.e. whether it is logically consistent.
4. The system has been developed with a view to being an aid to the social worker (i.e. it acts as a specialist social worker). However, the idea of an expert system being used directly by clients is an area which we believe needs active investigation.
5. Methods will need to be developed which can deal with non-rule-based knowledge and the seemingly illogical, yet common, occurrence of two separate philosophical positions being held which occasionally produce contradictions.

CONCLUSION

Although we have built only a small prototype expert system for one particular area of social work (*enuresis*) it has led us to believe that such systems could have a major impact in the field of social work in the future. By using an expert system shell it is possible to obtain a working prototype within a matter of days or a few weeks. However, the early results of our study led to the belief that certainly this low-cost expert system shell is only suitable for prototyping and is not suitable for a full working system. For this reason we are now developing a new system based upon ideas perceived during the building of the prototype.

REFERENCES

Expert systems

Buchanan, B. G. (1984). *Rule-based Expert Systems: The MYCIN Experiments*, Addison-Wesley.
Gaschnig, J. (1982). 'Prospector: an expert system for mineral exploration'. In *Introductory Readings in Expert Systems* (Ed. D. Michie), Gordon and Breach, pp. 47–64.
Hayes-Roth, F., *et al.* (1983). *Building Expert Systems*, Addison-Wesley.
McDermott, J. (1982). 'RI: a rule-based configurer of computer systems'. *Artificial Intelligence*, **19**, 39–88.
Nilsson, N. J. (1982). *Principles of Artificial Intelligence*, Springer-Verlag.
Shortliffe, E. H. (1976). *Computer-Based Medical Consultation: MYCIN*, Elsevier.
Waterman, D. (1985). *A Guide to Expert Systems*, Addison-Wesley.

Microexpert

Cox, P. R. (1984). 'How we built micro expert'. In *Expert Systems: Principles and Case Studies* (Ed. R. Forsyth), Chapman and Hall, pp. 112–132.
Cox, P. R., and Broughton, R. K. (1984). *Micro Expert: Users Manual Version 3.1*, ISIS Systems Ltd, Redhill, Surrey.

Computers in Social Work

Geiss, R. (1985). *Information Technology and Social Work*, Howard Press, New York.
Schoech, D., Jennings, H., and Schkade, L. H. R. C. (1985). 'Expert systems, artificial intelligence for professional decisions'. *Computers in Human Services*, **1.1**, 81.

Enuresis

Kolvin, R. C., and MacKeith (1973). *Bladder Control and Enuresis*, Ludan.
Morgan, R. T. T. (1974). *Enuresis and the Enuresis Alarm—A Clinical Manual for the Treatment of Nocturnal Enuresis*, Child Treatment Research Unit, School of Social Work, University of Leicester.
Toole, S. K. 'Incontinence in children'. Unpublished Manuscript.
Turner, R. K. (1973). 'Conditioning treatment of nocturnal enuresis: present status'. *Bladder Control and Enuresis* (Eds R. C. Kolvin *et al.*), Heinemann.
Turner, R. K., *et al.* (1975). *A Behavioural Approach to the Treatment of Diurnal Enuresis*, Child Treatment Research Unit, School of Social Work, University of Leicester.

Note: It is proposed to publish further articles on the subject in forthcoming editions of *Computer Application in Social Work*, CASW (Publications) Birmingham.

Artificial Intelligence for Society
Edited by K. S. Gill
© 1986 John Wiley & Sons Ltd

26.

LANGUAGE DEVELOPMENT AND INTERACTIVE TECHNOLOGY

FARHAD JAHEDI SEAKE Centre, Brighton Polytechnic

ABSTRACT

This paper presents the design of an interactive learning system for language development for children with learning difficulties. The system is primarily designed as a tool for the teacher to use it as an integral part of a language teaching programme. It also enables children to explore their own concepts of language development through experimentation.

Design of the system integrates three forms of representations, text, sound and graphics, which provide a learning environment that is more descriptive and expressive.

INTRODUCTION

Until recently, most of the microcomputer systems for education have been designed for the teaching of limited domains through a rigid system of success and failure. These systems reflect certain tutorial strategies based on knowledge extracted from a tutor, and the design is normaly based on mundane techniques such as those of drill and practice and 'scoring' dependent upon the success and failure rates of the student. Students find themselves in a learning environment in which they are constantly pushed towards or away from a predetermined goal. Freedom of choice under these systems belongs entirely to the computer and the student has hardly any role to play in his learning process.

As a result of recent developments in the use of computers in education, a number of new methodologies and approaches have been introduced. Much of software developed in this phase is associated with the familiar names: CAI (computer-aided instruction), CAL (computer-aided learning) and CBL (computer-based learning).

At the same time, growing interest in the applications of artificial intelligence (AI) in education has resulted in the emergence of new techniques and theories. Systems that have been developed in this area are generally known as ICAI (intelligent computer-aided instruction) or ITS (intelligent tutoring system). Systems of this type should be able to make decisions based on a generalized set of teaching rules. A decision made by the tutoring system should reflect the type of misunderstanding or mistakes made by the student. Rate of success or failure in these systems is not the basis for introducing new teaching material.

The system presented in this paper is part of the work of the Social and Educational Applications of Knowledge Engineering (SEAKE) Centre on the development of interactive learning tools for disadvantaged and handicapped children.

One of the main areas of concern has been the application of computer technology to enhance the social communication and life skills of children with learning difficulties. Language development is seen as the medium through which the skills can be enhanced and contribute to development of general human competences. The emphasis of the work is on creativity, expressiveness, learning through experimentation and sharing of experiences both inside and outside the classroom situation. Initially a number of computer programs were developed in the area of computer animation, story generation, computer music and speech to help handicapped children in their communication and control within and of their environments.

Recent work has included educational games for reading and writing, and prototype projects on interactive video for social communication and life skills. It was felt both by the SEAKE Centre team and the teachers from our local special schools that there was a need to design a microcomputer-based system which enables the teacher to use the computer facilities of graphics, sound and text in various ways for generating lessons to meet the children's individual as well as group language development needs. The system in question has been designed to meet these criteria as well as to provide a basis for further work into intelligent knowledge-based learning aids for disadvantaged children.

INTERACTIVE LEARNING AIDS FOR LANGUAGE DEVELOPMENT

The design of interactive learning aids for language development should take into account the learning needs of the child, the curriculum needs of the school and the social needs of the community within which the child and the teacher interact. In this context, we consider the design of computer aids for language development as central to our work on social communication and life skills.

Learning is a long and continuous process that occurs throughout life. One of the significant factors for learning is motivation and the need to learn. Such factors may themselves be dependent upon other factors such as learning environment, life experiences, goals and social factors, and physical constraints. Children with learning difficulties often lack self-confidence and motivation. Controlled learning environments often reduce the child's ability to become independent and to gain confidence. Some children with learning difficulties also suffer from physical disabilities which affect their communication abilities. For example, dyslexic children, deaf children and children with speech impairments often have serious difficulties in reading and writing. Their limited contact with the spoken form of language is often the reason for their lack of knowledge about the language. This problem is often exacerbated by the lack of proper equipment in their schools. One important factor which must also be considered in teaching mentally retarded children is that they require individualized attention since they have different learning needs, ability and skills. However, the cost of providing this individual attention is considered to be too high and hence the needs of the child are not met; consequently the child's capabilities are not enhanced.

Children with moderate learning difficulties who may also be mentally retarded suffer acutely from the lack of motivation and autonomy. This results in a loss of interest and inability to concentrate for long periods. To overcome this lack of motivation, a computer-aided teaching/learning tool should be designed so as to take into consideration the child's own interests and skills. This means that a learning situation should be made more explicit and its relationships and associations with the child's world must be demonstrated. Graphics, animation and in general visual forms of representations can prove to be very important for creating motivating environments.

Some of the cognitive and linguistic issues that affect learning processes can be identified as:

Problem solving
Sequencing
Organizing
Analysing
Reasoning
Explaining
Deducing

The above in turn form a part of larger problem areas such as: short- or long-term memory problems, limited vocabulary, and abstraction and generalization of knowledge. However, to create an interactive learning environment the following skills of the learner must be enhanced:

1. *Listening* requires an appropriate use of audio sensors in capturing the words in isolation as well as in context. Diversion of attention from this may prove fatal to the learning process.
2. *Talking*. This skill develops with improved listening. Extensive vocabulary, reasoning and explaining abilities are essential elements of good speech.
3. *Reading*. Visual skills of differentiation, recall and interpretation of the written material, comprehension and analysis of the written material are the types of skills needed for developing reading skills.
4. *Writing*. Expanded vocabulary, ability to express, ability to formulate and sequence situations recalled from memory, knowledge of rules of syntax are some of the skills needed to develop writing.

DESIGN ISSUES

Current research indicates that theories of open learning environments and exploration are being widely accepted in education and are proving to be very effective. AI researchers such as Papert,[1] Sleeman and Brown[2] and O'Shea and Self[3] have made major contributions in this area.

AI environments for education such as LOGO[1] and MicroPROLOG[4] have made AI languages available at school level. AI systems such as intelligent tutoring systems and expert systems have developed techniques such as the user modelling and bug finding. Rule-based systems provide a significant resource for designing intelligent learning environments for education. From these developments, it has become apparent that any subsequent work in the area of computer in education must include an adequate system of user modelling. User modelling reflects the user's ability to learn a concept and to solve problems in the context of a given tutorial.[5] This may be achieved by keeping the student's performance history and by monitoring his/her performance.

One of the advantages of the new technology is in the programmability of the computer. It can be programmed to perform difficult tasks and to achieve a predetermined goal. It is, however, important that the goal be clearly identified, that the necessary interaction between the user and the machine be built into the system and that the design of the system be oriented towards the user. Very often inadequacies in the use of the microcomputers result in programming the child rather than the computer. This has the effect of reducing the child's expressive power as well as his/her motivation.

Communication is an important aspect of any learning environment. When a student is unable to communicate through conventional methods, that is to say, when he/she cannot speak or write, other forms of communication such as Bliss Symbol or a sign language may be used.

Today's computers are flexible and offer a wide range of interfaces that make the user interaction with the machine easy. Touch-sensitive screens, special switches, joysticks, speech input (recognition), speech output (synthesizers, digitizers), graphic pads and screens, interactive videos and text screens are some of these facilities. When designing computer programs for teaching the facilities provided by a microcomputer should be used to the full potential so as to produce a user friendly interactive system.

User friendliness requires humour and a human touch that helps to reduce the formal and off-putting nature of the machine. The error-trapping routines should not deter or distract the user's attention from the continuous flow of the main program. Messages should be gentle but meaningful and in no way should the user be pressed for time unnecessarily. Visual and auditory means of communication should be used to express messages as necessary. Frequent help-tutorial may be offered in cases where the user is expected to have serious difficulties of interaction.

DESIGN CRITERIA

In the light of the above discussion, we consider the following criteria for the design of the system:

1. A non-restricted learning environment where the user is not pushed towards a predetermined goal, but is motivated towards achieving the goal.
2. Monitoring of the student's performance in order to establish his/her level of knowledge and confidence in carrying out tasks that are relevant to the processes of language development.
3. Feedback material that is prepared by the teacher for providing remedial help to the student.
4. A one-to-one interaction in a relaxed environment.
5. Tools which the student can use to combine animation, music and text in the form of a scenario which reflects skills and experiences. This in turn forms the basis of an analysis of the student's learning abilities.

As a useful tool in the classroom, the computer should play the part of an intelligent companion that is ever patient, friendly and understanding.

A prototype working program (INLEARN) has been developed. It integrates the three forms of representations, visual, auditory and written, which form the main body of the system's knowledge base. It allows the user full access to its facilities. These include:

1. A *shape manager* which allows the user to draw shapes, animate these shapes and store the descriptions of the objects in a textual form.

2. A *music composer* which enables the user to create and edit tunes and write lyrics for the stored tunes.

   INLEARN generates 'help' either on user's request or by detection. A current error-trapping routine detects errors only when the cumulative number of detected errors exceeds that of a predetermined number. Tutorial sessions are often offered as a result of user's request.

3. The *experimental mode* allows the user to manipulate shapes and tunes on the VDU's screen by instructing the machine through simple sentences or phrases.

Throughout interaction with INLEARN, the user's activities are monitored and kept in a separate file. This file represents the user's strategies as well as mistakes.

The INLEARN package includes a booklet which contains the necessary information, sample sessions, examples and exercises.

The package provides a learning environment in which the text, graphics and sound are integrated to provide a medium for representing the various concepts of language development.

CONCLUSION

The first phase of the project is accomplished and, as a result, a microcomputer package has been developed which enables the user to integrate text, graphics and music in various ways.

Children can use simple words, sentences and phrases to create their own animation, compose their favourite music, write lyrics for thier composed tunes. Their own knowledge can be extended by solving problems associated with examples given to them in the booklet accompanying the package.

The teacher can use the package to design teaching material which contributes to enhancing learning capabilities of the individual child as well as the group.

However, the main drawback in such a system is the limited memory size of the computers which are available in schools as well as the lack of AI languages and software tools for these computers. It is intended to use the pilot program in a number of special schools with a view to designing an intelligent knowledge-based system for language development. It is also proposed to use an interactive videodisc as part of a multimedia learning environment for children with special needs. Interactive video provides a powerful source for visual knowledge of realistic and relevant social and cultural contexts within which language development takes place. AI techniques such as user modelling could be used to study the pupil–machine interaction with a view to providing relevant remedial assistance and guidance which is meaningful to the pupil with the given contexts mentioned above. To achieve these aims, the SEAKE Centre has initiated projects on IKBS/interactive video for social communication and life skills for disadvantaged youths and adults.

REFERENCES

1. Papert, S. (1980). *Mindstorms: Children, Computers and Powerful Ideas*, Harvester Press.
2. Sleeman, D., and Brown, J. S. (1983). *Intelligent Tutoring Systems*, Academic Press.
3. O'Shea, T., and Self, J. (1983). *Learning and Teaching with Computers*, Harvester Press.
4. Clark, K. L., and McCabe, F. G. (1984). *MicroPROLOG: Programming in Logic*, Prentice-Hall International.
5. Jahedi, F., and Jones, S. G. (1985). 'Intelligent tutoring systems'. Working Paper No. W-201, SEAKE Centre, Brighton Polytechnic.

BIBLIOGRAPHY

Chambers, J. A., and Sprecher, J. W. (1983). *Computer Assisted Instruction*, Prentice-Hall.
Chandler, D., and Marcus, S. (1985). *Computers and Literacy*, Open University Press.
Goldenberg, E. P. (1979). *Special Technology for Special Children*, University Park Press.
Goldenberg, E. P., Russell, S. J., Carter, C. J., Stokes, S., Sylvester, M. J., and Keiman, P. (1984). *Computer, Education and Special Needs*, Addison-Wesley.
Hartley, J. R., and Sleeman, D. (1983). 'Towards more intelligent teaching systems'. *International Journal of Man–Machine Studies*, **5**, 215–236.
Kahn, K. (1977). *Three Interactions between AI and Education in 'Machine Intelligence 8'*, Ellis Harwood.
O'Shea, T., Bornet, B., Du Boulay, B., Eisenstadt, M., and Page, I. (1981). *Tools for Creating Intelligent Computer Tutors*, NATO Symposium on Human and AI.

Artificial Intelligence for Society
Edited by K. S. Gill
© 1986 John Wiley & Sons Ltd

27.

AUDIO-VISUAL ARTIFICIAL COMMUNICATION

Z. M. ALBES Principal Psychologist, Eastbourne, East Sussex

I will speak of the role and value of the latest development through communication aids which come in the category of artificial communication. I will point out the psychological needs, the need for integration, the burden of family commitment, frustration and stress factors for patient and family, supporting cooperation of speech therapists, methods of self-therapy, help in communication for autistic, cerebral palsy, spastic or Down's syndrome children, as well as the speech-impaired adult, especially the stroke victim; the importance and value of microchip technology; the need of flexibility to use programs as flexible as possible to the needs of the individual, taking into consideration their mental and physical ability.

I have chosen to speak about the role and value of the latest technological developments through artificial communication aids in modern medicine and education as a therapeutic support and as an educational one. I will point out the psychological facts and the needs for integration of the non-speaking—the speech-impaired— person in our society. In order to enable his integration, the professionals, whether they are psychiatrists, psychologists, neurologists or speech therapists, need to enlist the help of the whole family; they cannot do it by themselves.

A handicapped person in the family concerns the whole family; it is not only the handicapped person who suffers from frustration, tension and stress. In fact, in some cases the burden is heavier to bear by a wife, husband or parent, than actually by the patient himself or herself. On the one hand you have the patient who cannot communicate, cannot express thoughts or feelings—the psychological consequences are shown to you all. On the other hand you have the members of the family who feel frustrated by not being able to help.

These are the facts we have to deal with. Over the duration of thirty years of my involvement with all kinds of handicapped, I have searched for ways to help. As a psychologist my aim was to find a way to reduce their frustration and many other psychological consequences—to replace natural communication, if damaged, with artificial communication (the use of the human voice to reduce impressions of artificiality, so as to enable the patient to identify himself with the sound of the spoken word).

In the United Kingdom alone we have about two hundred thousand speech-impaired persons; in America the number is over one and a half million. This is an enormous number of people who are being condemned to live in mental isolation, never to be able to express their thoughts or their feelings. For these reasons I became inventive and I felt a great urge to help this group so near to my heart.

Even in the hospitals the speech therapists and occupational therapists cannot usually devote more than a half an hour once or twice a week per patient. This is certainly not a sufficient amount of time to retrain speech in order to regain speech. Here my methods have been most valuable as I have developed a logical symbol system by using visual messages, especially useful for those patients where reading and memorizing sentences, even words, are difficult. 'Convaid' is one of those aids which provides artificial communication; it is not only the means of communication but stores the words in sequences and forms sentences. This is a relief for the patients as, in many cases, the greatest problem is that although they know what they would like to say, they can only say one word or part of a word. They cannot memorize the whole sentence. By using my method we have achieved results where the patient can train himself/herself daily with or without the assistance of the family. By using earphones, the patient can simultaneously repeat the words correctly offered. During our research we found that a patient who had not spoken for eleven years made excellent progress by using this method.

Those who have worked with autistic children, with cerebral palsy children, the spastic or the Down's syndrome child will understand my own frustration of not being able to help. I wanted desperately to find a way and this communication aid is the result of my determination.

I would not, though, wish to appear to be diminishing the value of any other existing speech aids. They are an enormous help. They are in a way the most valuable contribution of our day, made possible only through microchips. Some of them are highly technical and sophisticated; others need a certain intelligence to be used, even understood; some are huge; most of them need good motor skill control to be able to be operated by finger; most of them need typing skills, spelling and reading ability.

Although the names of existing speech aids will not be mentioned here, all of them have their purpose and many are a great help. Some of them have limitations such as their complexity, which require a higher IQ and motor skills to operate. Some of them are more or less only for simple conversation. I tried to find a solution where the machine can be adapted to the needs of the patient instead of the patient having to adapt to the limitations of the machine.

At this point, before I introduce my own invention, 'Convaid' (which has been developed at the Bio-Medical Engineering Department of the University of Sussex by Dr Barrie Jones and Dr Jeremy Watson), let me make it clear that artificial intelligence is and what artificial communication aids are. Then let us analyse controversial opinions as to how useful or how damaging such an approach can be.

Consider the case where a heart, an arm or a leg has been damaged and is, therefore, not able to carry out its function. In such a case, if it can be replaced partly or wholly by artificial means, we should not hesitate to do it. Naturally there are cases where the artificial approach is not needed indefinitely and it would be wrong to use it and therefore weaken the chances of regaining natural functions. However, before going any further or deeper into this subject we must establish that even if we use artificial means it is important to use, as much as possible, natural functions which can and must be developed.

My methods are developed according to the patient's own individual needs. The programmes are not standardized and they can be provided with a vocabulary chosen by the family as they require it. It is interchangeable for each situation, IQ, social, medical, educational and therapeutic needs. Because of this the vocabulary has no limitations. Furthermore, any language can be chosen, even a bilingual one, which means you can communicate with your patients in any language without an interpreter. This in my judgement is an important and a great asset, especially in the psychiatric field where communication is of the greatest importance. Every doctor, but mainly every psychiatrist, should have such a facility as any craftsman needs his tools. A communication aid of this type should be the psychiatrist's tool. I am speaking from my own experience and the results of our survey which we carried out during a year of field research in the United Kingdom.

We have involved 69 patients, held seminars for 180 therapists and have demonstrated 'Convaid' at three exhibitions where we have been visited by hundreds of professionals and handicapped people. The impression made by this aid has exceeded our expectations.

I actually started to use my speech aid with an autistic child and then with a spastic child. As a child psychologist my natural interest was primarily the child, but when we started our field research and visited several hospitals in London and other counties, doctors of several disciplines approached us and asked us to try our speech aid with their patients. To my great satisfaction, 'Convaid' proved to be very successful. It worked for stroke patients and, in some cases, for those suffering from multiple sclerosis, transient loss of speech after shock, motor-neurone disease, Parkinson's disease and many others including aphomia and aphrosia.

The following points need to be taken into considerations.

THE TWO MAIN FUNCTIONS OF 'CONVAID'

1. It acts to open, process and store information received from the surroundings.
2. It is a unit for programming, regulating and varying mental action. In many cases where these functions are impaired, this aid has proved to be supportive for these basic necessities.

AREAS ASSOCIATED WITH SPEECH PROCESSING AND THEIR INTERCONNECTION

Broca's Area

Here it provides control signals to the physiological speech-production mechanism—vocal chords, larynx, etc.—and accepts speech data via the arcuatic fasciculus from Wernicke's area.

Wernicke's Area

Located close to the region of the cortex which receives auditory stimuli, Wernicke's area handles the cognitive processing of speech patterns, including their association with concepts.

Angular Gyrus

This links the visual cortex with Wernicke's area, allowing the spoken interpretation of visual linguistic patterns, e.g. text; the reverse operation is also likely, whereby auditory speech patterns evoke visual symbols, as in writing.

Functional (block) diagrammatic representation

Concept	Wernicke's area cognition	Broca's area coordinated	Vocal tract neuromuscular system	(Speech)
	Malfunctions here produce aphasia/ dyphasia.	Malfunctions here produce dysarthia.		

AREAS IN THE BRAIN FOR WHICH 'CONVAID' MAY PROVIDE PROSTHETIC SUPPORT

The main area in the brain whose malfunction may be compensated for by the prescription of 'Convaid' is Broca's area. Assuming Wernicke's area is intact, the patient should be able to associate abstract concepts with internal auditory and visual representation, thereby being able to access utterances on the aid by symbols. Disabilities involving damage to the neuromuscular subsystems responsible for the articulation of the speech-production mechanisms (such as found with muscular sclerosis, etc.) may also be alleviated by the use of 'Convaid'.

In certain cases (e.g. following some strokes) Wernicke's area may be partially damaged. In such a case, the difficulty is found to be in associating abstract concepts with verbal auditory representations. Where an adequate path to the angular gyrus exists, visual symbols may be recognized as representing concepts requiring expression. In these cases, 'Convaid' can provide a functional link 'across' Wernicke's area and if a personal earphone is used it can 'prompt' the speech-impaired user. In children, this mode of operation may encourage plasticity in the neural networks and facilitate other, undamaged, areas of the cortex by taking over parts of the speech-processing role.

In cases of aphasia, when speech disturbance occurs due to illness, if the central speech apparatus stays intact as well as the intelligence but the patient suffers from a lack of capability to transfer meanings to words and to comprehend written words, our speech aid is a vital support. It helps the patient who is unable to have spontaneous speech and who has delay in repeating words formed in the mind. In cases where additional paralexia is also evident, the patient has difficulty in reading as the words seem to him to be confused; 'Convaid' is especially useful in such a case as it represents the written word in drawn symbols. In cases of amnesia where words cause difficulties, if the word is correctly offered it will be recognized and repeated without mistake. Here again the aid is invaluable as it provides the correct word, which by recognition through the symbols can then be correctly repeated. For this very reason the aid can be used not only for communication but also as an exercise.

In the case of aphomia—lack of voice due to larynx damage or only having a very whispering voice—'Convaid' is the ideal communication aid.

In the case of aphrosia, where the buildup of the sentence is impossible, our aid is valuable as it builds up the sentence in sequences from single words chosen by the patient, whose frustration lies in not being able to organize thoughts. The patient can use the aid to take over this task, storing messages given by the user and so expressing these in an uncomplicated spontaneous way.

The frustration and tension is reduced to a minimum as the patient becomes more and more familiar with my methods and the use of 'Convaid'. The patient can interchange unlimited programmes to his own liking according to his environment and state of intelligence. Hence, as can be seen from the applications described throughout this paper, it is very clear that there are enormous psychological benefits to be derived from using speech aids.

Index